CHROMATOGRAPHY: CONCEPTS, METHODS AND APPLICATIONS

Edited by
Levitah Castil Mapatac

Delve Publishing

Chromatography: Concepts, Methods and Applications

Edited by Levitah Castil Mapatac

Edition 2018

Delve Publishing

Published by
Delve Publishing LLC
708 3rd Avenue, 6th Floor
New York
NY 10801
United States of America
www.delvepublishing.com

ISBN: 978-1-68095-722-8 (Hardback)

Printed in United States of America

Library of Congress Control Number: 2016920402

About the Editor

Levitah Castil Mapatac

Levitah C. Mapatac is an Associate Professor, researcher, writer and a PhD holder from the University in Science and Technology of Southern Philippines. Her interest varies from different chemistry concepts and applications and currently doing research in traditional medicines, water waste pollution and environmental chemistry in the aspect of contamination pathways, profilling and rehabilatation of mining waste.

Table of Contents

List of Abbreviations xiii
Preface xvii

1 INTRODUCTION TO CHROMATOGRAPHY

Introduction 1
History 3
Chromatography Terms 3
Displacement Chromatography 5
Techniques by Separation Mechanism 6
    Ion Exchange Chromatography 6
    Size-Exclusion Chromatography 6
    Expanded Bed Adsorption Chromatographic Separation 6
References 7

2 DISTRIBUTION CONSTANT

Introduction 9
    Expression 10
Freundlich Equation 10
    Freundlich Adsorption Isotherm 10
    Limitation of Freundlich Adsorption Isotherm 11
Kovats Retention Index 11
    Expression 12
Retardation Factor 12
    General Definition 13
    Planar Chromatography 13
Van Deemter Equation 14
    Van Deemter equation 15
    Expanded Van Deemter 17
References 17

3 TECHNIQUES BY CHROMATOGRAPHIC BED SHAPE

Introduction 19
- Column Chromatography 20
- Planar Chromatography 26
- Paper Chromatography 26
- Thin Layer Chromatography (TLC) 27

High-Performance Liquid Chromatography 34

Operation 36
- History and Development 37
- Types 38
- Normal–Phase Chromatography 39
- Displacement Chromatography 40
- Reversed-Phase Chromatography (RPC) 41
- Size-Exclusion Chromatography 43
- Ion-exchange Chromatography 43
- Bioaffinity Chromatography 44
- Aqueous Normal-Phase Chromatography 44

Isocratic and Gradient Elution 44

Parameters 45
- Theoretical 45
- Internal Diameter 46
- Particle Size 47
- Pore Size 48
- Pump Pressure 48
- Detectors 48
- Autosamplers 48

Applications 48
- Manufacturing 48
- Legal 49
- Research 49
- Medical 49

Fast Protein Liquid Chromatography 49
- FPLC System Components 50
- FPLC Columns 52
- Optimizing Protein Purification 52

References 53

4 GAS CHROMATOGRAPHY–MASS SPECTROMETRY

History 58

Instrumentation 59

Purge and Trap GC-MS 60
Types of Mass Spectrometer Detectors 60
GC-tandem MS 61
Ionization 61
Electron Ionization 61
Cold electron Ionization 62
Chemical Ionization 62
Analysis 62
Full scan MS 63
Selected Ion Monitoring 64
Applications 64
Environmental Monitoring and Cleanup 64
Criminal Forensics 64
Law Enforcement 64
Sports Anti-Doping Analysis 64
Security 64
Chemical Warfare Agent Detection 65
Chemical Engineering 65
Food, Bbeverage and Perfume Analysis 65
Astrochemistry 65
Medicine 66
Liquid Chromatography–Mass Spectrometry 66
Liquid Chromatography 66
Mass Spectrometry 67
Applications 68
Pyrolysis–Gas Chromatography–Mass Spectrometry 69
How it Works 69
Applications 69
Affinity Chromatography 70
Uses 70
Principle 70
Batch and Column Setups 70
Specific Uses 72
Weak Affinity Chromatography 75
Protein Targets 75
Advantages 75
References 76

5 CAPILLARY ELECTROCHROMATOGRAPHY

Introduction 80
Principle 80

Instrumentation and Working 81
Advantages 82
References 82

6 SEPARATION MECHANISM TECHNIQUES

Ion Chromatography 83
History 85
Principle 86
Procedure 87
Weak and Strong Ion Exchangers 89
Typical Technique 90
Membrane Exchange Chromatography 91
Separating Proteins 91
    Gibbs-Donnan Effect 92
Uses 93
    Clinical Utility 93
    Industrial Applications 94
    Drug Development 94
Size-Exclusion Chromatography 95
    Applications 95
    Advantages 95
    Discovery 96
    Theory and Method 96
    Factors Affecting Filtration 98
    Analysis 99
    Applications 100
Expanded Bed Adsorption 102
    Principle 102
References 103

7 SPECIAL TECHNIQUES OF CHROMATOGRAPHY

Reversed-Phase Chromatography 109
Stationary Phases 110
    Silica-based Stationary Phases 110
Mobile Phases 111
Simulated Moving Bed 111
    Construction 112
    Advantages 112
    Drawbacks 112
    Applications 113

Countercurrent Chromatography 113
History 114
Support-free Liquid Chromatography 114
Droplet CCC 116
Hydrodynamic CCC 117
Hydrostatic 118
Modes of Operation 119
Applications 124
References 124

8 CHROMATOGRAPHY DETECTORS

Introduction 133
Destructive Detect 134
Non-Destructictors 134
Flame Ionization Detector 135
History 135
Operrinciple 135
Advantages and disadvantages 136
Alternative Solution 136
Operation 136
Description of a Generic Detector 137
Nitrogen–Phosphorus Detector 137
Evaporative Light Scattering Detector (ELSD) 138
Principles of Operation 138
Mass Spectrometry 138
Parts of a Mass Spectrometer 141
Creating ions 142
Inductively Coupled Plasma 144
Sector Instruments 146
Detectors 149
Tandem Mass Spectrometry 150
Data and Analysis 153
Interpretation of Mass Spectra 154
Applications 155
References 159

9 ELECTROCHROMATOGRAPHY

Introduction 165
Micellar Electrokinetic Chromatography 166
Theory 166

History 168
Applications of Mekc 168
Displacement Chromatography 168
Discovery 169
Principle 169
Mode of Operation 170
Comparison with Elution Chromatography 171
Applications 172
References 173

10 CAPILLARY ELECTROPHORESIS

Introduction 177
Instrumentation 178
Detection 179
Modes of Separation 180
Efficiency and Resolution 183
Applications 184
Capillary Electrophoresis–Mass Spectrometry 185
Interfacing CE with MS 185
Electrospray Ionization Interface 186
Applications 188
Tandem Mass Spectrometry 189
Instrumentation 189
Tandem in Space 190
Tandem in time 190
Negative Electron Transfer Dissociation 193
Electron-Detachment Dissociation 194
Charge Transfer Dissociation 194
Tandem Mass Tag 196
Applications 196
Electrostatic Precipitator 197
Invention of the Electrostatic Precipitator 198
Plate Precipitator 199
Collection Efficiency (*R*) 200
Low Resistivity 206
Modern Industrial Electrostatic Precipitators 210
Electrostatic Air Sampling 210
Wet Electrostatic Precipitator 211
Consumer-Oriented Electrostatic Air Cleaners 211
Microfiltration 212
General Principles 212

Range of Applications ... 213
Main Process Characteristics ... 214
Process and Equipment Design ... 215
Membrane Equipment ... 218
Permeate Flux ... 219
Environmental Issues, Safety and Regulation ... 220
Comparison with Similar Processes ... 221
Recent Developments ... 221
References ... 222

11 GEL PERMEATION CHROMATOGRAPHY

Introduction ... 227
How GPC Works ... 228
Application ... 229
Material and Methods ... 230
Instrumentation ... 230
Gel ... 231
Column ... 231
Eluent ... 231
Pump ... 231
Detector ... 231
Data Analysis ... 232
Advantages of GPC ... 234
Disadvantages of GPC ... 234
Orthogonal Methods ... 234
References ... 235

Index ... 241

List of Abbreviations

ASEC	Absolute Size-Exclusion Chromatography
NQA	Aerosol-Based Detector
ANP	Aqueous Normal-Phase Chromatography
APCI	Atmospheric-Pressure Chemical Ionization
AED	Atomic-Emission Detector B
BSA	Bovine Serum Albumin
CAE	Capillary Array Electrophoresis
CEC	Capillary Electrochromatography
CE	Capillary Electrophoresis,
CGE	Capillary Gel Electrophoresis
CIEF	Capillary Isoelectric Focusing
CZE	Capillary Zone Electrophoresis
CPC	Centrifugal Partition Chromatography
CPE	Centrifugal Partition Extraction
CTD	Charge Transfer Dissociation
CI	Chemical Ionization
CWA Q	Chemical Warfare Agents
CIP	Cleaning-in-Place
CID Q	Collision-Induced Dissociation
DRI	Differential Refractometer
DCCC	Droplet Counter Current Chromatography
ECD	Electron Capture Detector
EDD	Electron-Detachment Dissociation
EAI	Electronic Associates, Inc.

EI	Electron Ionization,
ESI	Electrospray Ionization
ESP	Electrostatic Precipitator
ELSD	Evaporative Light Scattering Detector
EBA	Expanded Bed Adsorption
FAB	Fast Atom Bombardment
FPLC	Fast Protein Liquid Chromatography
FFF	Field-Flow Fractionation
FID	Flame Ionization Detector
FPD	Flame Photometric Detector
GC	Gas Chromatography
GPC	Gel Permeation Chromatography
HPLC	High-performance Liquid Chromatography
HIC	Hydrophobic Interaction Chromatography
IC	In Ion-Exchange Chromatography
LALLS	Light Scattering Detectors
LC	Lliquid Chromatography
MS	Mass Spectrometer
MD	Mira Detector
MAL	Mobile Analytical Laboratories
MCSGP	Multicolumn Countercurrent Solvent Gradient Purification
MRM	Multiple Reaction Monitoring
NCI	Negative Chemical Ionization
ONP	Organic Normal Phase Chromatography
PID	Photoionization Detector
PDI	Polydispersity Index
PTT	Poly(Trimethylene Terephthalate)
PCI	Positive Chemical Ionization
PTV	Programmable Temperature Vaporizer
RI	Refractive Index,
RP	Reversed-Phase Chromatography
SIM	Selected Ion Monitoring

SRM	Selected Reaction Monitoring
SMB	Simulated Moving Bed
SEC	Size-Exclusion Chromatography
SDS	Sodium Dodecyl Sulfate
SLD	Soft Laser Desorption
SPME	Solid-Phase Micro Extraction
SMSD	Sumon Detector
SMB	Supersonic Molecular Beam
TMT	Tandem Mass Tag
THF	Tetrahydrofuran
TCD	Thermal Conductivity Detector
TLC	Thin Layer Chromatography
TMBC	True Moving Bed Chromatography

Preface

Chromatography has become a cornerstone of separation science, that branch of chemistry devoted to separating compounds from mixtures. Chromatography is one of several separation techniques defined as differential migration from a narrow initial zone. Electrophoresis is another member of this group. In this case, the driving force is an electric field, which exerts different forces on solutes of different ionic charge. The resistive force is the viscosity of the nonflowing solvent. The combination of these forces yields ion mobilities peculiar to each solute. There are two main categories of chromatography: preparative and analytical. Analytical work uses small sample sizes; the objective is to separate compounds in order to identify them. Preparative work uses large quantities of samples and collects the output in bulk; the point of the chromatography here is to remove impurities from a commercial product. The two are not mutually exclusive. Chromatography has numerous applications in biological and chemical fields. It is widely used in biochemical research for the separation and identification of chemical compounds of biological origin. In the petroleum industry the technique is employed to analyze complex mixtures of hydrocarbons. Pharmaceutical industries use this method both to prepare huge quantities of extremely pure materials, and also to analyze the purified compounds for trace contaminants. These separation techniques like chromatography gain importance in different kinds of companies, different departments like Fuel Industry, biotechnology, biochemical processes, and forensic science. Chromatography is used for quality analyses and checker in the food industry, by identifying and separating, analyzing additives, vitamins, preservatives, proteins, and amino acids. Chromatography like HPLC is used in DNA fingerprinting and bioinformatics.

Content Coverage

Chapter One presents the introduction of chromatography. Chromatography technique developed substantially as a result of the work of Archer John Porter Martin and Richard Laurence Millington Synge during the 1940s and 1950s. Since then, the technology has advanced rapidly.

Chapter Two focuses on the distribution constant, which is the concentration of a component in or on the stationary phase divided by the concentration of the component in the mobile phase in equilibrium conditions.

Chapter Three outlines the different kinds of chromatographic techniques and these are classified according to the shape of bed, physical state of mobile phase, separation mechanisms.

Chapter Four highlights about gas chromatography–mass spectrometry method that combines the features of gas-chromatography and mass spectrometry to identify different substances within a test sample..

Chapter Five focuses on capillary electrochromatography, an emerging analytical technique, which combines the versatility of chromatography with the high resolution of electrophoresis. The separation of complex analyte mixtures occurs as the result of a complex interplay between chromatographic retention and differential migration in the applied electric field.

Chapter Six emphasizes on separation mechanism techniques that depends upon differences in polarity between the different feed components. The more polar a molecule, the more strongly it will be adsorbed by a polar stationary phase. Similarly, the more non-polar a molecule, the more strongly it will be adsorbed by non-polar stationary phase.

Chapter Seven focuses on special techniques of chromatography such as reversed-phase chromatography that includes any chromatographic method that uses a hydrophobic stationary phase.

Chapter Eight presents an approach on chromatographic detectors capable of establishing both the identity and concentration of eluting components in the mobile phase stream. A broad range of detectors is available to meet different sample requirements.

Chapter Nine explores the electrochromatography separation technique used to resolve and separate mostly large biomolecules such as proteins. It is a combination of size exclusion chromatography (gel filtration chromatography) and gel electrophoresis.

Chapter Ten is about capillary electrophoresis that separates ions based on their electrophoretic mobility with the use of an applied voltage.

Chapter Eleven focuses on gel permeation chromatography that separates dissolved macromolecules by size based on their elution from columns filled with a porous gel.

CHAPTER 1

Introduction to Chromatography

Chapter Contents

- **Introduction**
- **History**
- **Chromatography Terms**
- **Displacement Chromatography**
- **Techniques by Separation Mechanism**

Chromatography is a process that can be used to isolate the various components of a mixture. There are a number of different types in use, including gas, liquid, paper, and gel permeation chromatography, and this process can get quite involved, especially with complex mixtures. It is also an extremely useful addition to a variety of fields, including pure and applied sciences, forensics, and athletics, among others.

INTRODUCTION

Chromatography is a laboratory technique for the separation of a mixture. The mixture is dissolved in a fluid called the *mobile phase,* which carries it through a structure holding another material called the *stationary phase.* The various constituents of the mixture travel at different speeds, causing them to separate. The separation is based on differential partitioning between the mobile and stationary phases. Subtle differences in a compound's partition coefficient result in differential retention on the stationary phase and thus changing the separation.

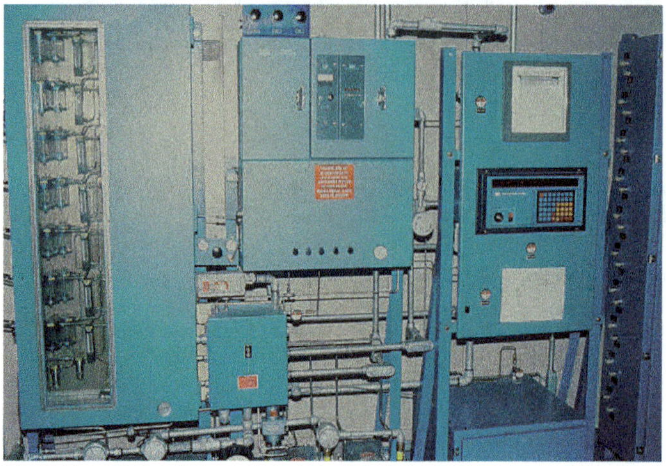

Pictured is a sophisticated gas chromatography system. This instrument records concentrations of acrylonitrile in the air at various points throughout the chemical laboratory.

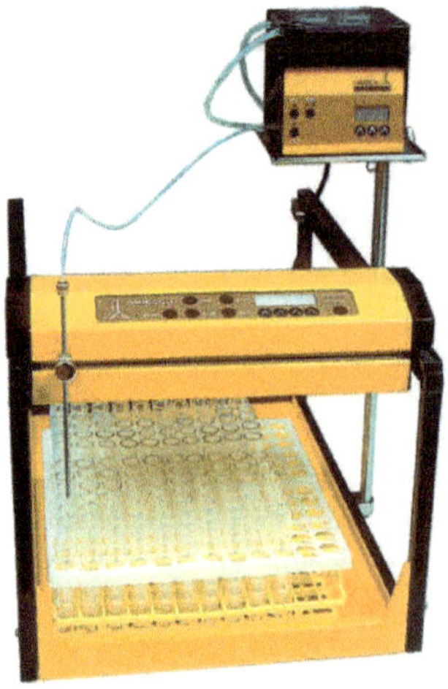

Automated fraction collector and sampler for chromatographic techniques.

Chromatography may be preparative or analytical. The purpose of preparative chromatography is to separate the components of a mixture for more advanced use (and is thus a form of purification). Analytical chromatography is done normally with smaller amounts of material and is for measuring the relative proportions of analytes in a mixture. The two are not mutually exclusive.

HISTORY

Thin layer chromatography is used to separate components of a plant extract, illustrating the experiment with plant pigments that gave chromatography its name.

Chromatography was first employed in Russia by the Italian-born scientist Mikhail Tsvet in 1900. He continued to work with chromatography in the first decade of the 20th century, primarily for the separation of plant pigments such as chlorophyll, carotenes, and xanthophylls. Since these components have different colors (green, orange, and yellow, respectively) they gave the technique its name. New types of chromatography developed during the 1930s and 1940s made the technique useful for many separation processes.

Chromatography technique developed substantially as a result of the work of Archer John Porter Martin and Richard Laurence Millington Synge during the 1940s and 1950s, for which they won a Nobel prize. They established the principles and basic techniques of partition chromatography, and their work encouraged the rapid development of several chromatographic methods: paper chromatography, gas chromatography, and what would become known as high performance liquid chromatography. Since then, the technology has advanced rapidly. Researchers found that the main principles of Tsvet`s chromatography could be applied in many different ways, resulting in the different varieties of chromatography described below. Advances are continually improving the technical performance of chromatography, allowing the separation of increasingly similar molecules.

CHROMATOGRAPHY TERMS

- The **analyte** is the substance to be separated during chromatography. It is also normally what is needed from the mixture.
- **Analytical chromatography** is used to determine the existence and possibly also the concentration of analyte(s) in a sample.

- A **bonded phase** is a stationary phase that is covalently bonded to the support particles or to the inside wall of the column tubing.
- A **chromatogram** is the visual output of the chromatograph. In the case of an optimal separation, different peaks or patterns on the chromatogram correspond to different components of the separated mixture.

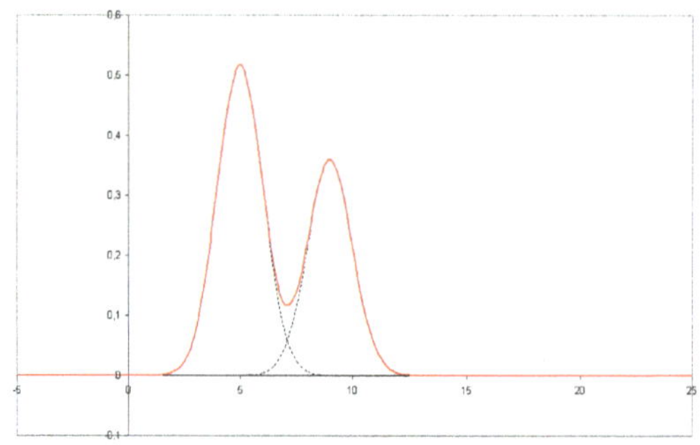

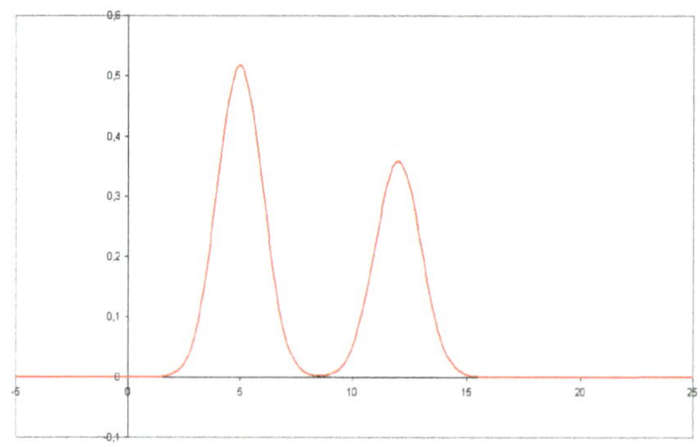

Plotted on the x-axis is the retention time and plotted on the y-axis a signal (for example obtained by a spectrophotometer, mass spectrometer or a variety of other detectors) corresponding to the response created by the analytes exiting the system. In the case of an optimal system the signal is proportional to the concentration of the specific analyte separated.

- A **chromatograph** is equipment that enables a sophisticated separation, e.g. gas chromatographic or liquid chromatographic separation.
- **Chromatography** is a physical method of separation that distributes components to separate between two phases, one stationary (stationary phase), the other (the mobile phase) moving in a definite direction.
- The **eluate** is the mobile phase leaving the column.
- The **eluent** is the solvent that carries the analyte.
- An **eluotropic series** is a list of solvents ranked according to their eluting power.
- An **immobilized phase** is a stationary phase that is immobilized on the support particles, or on the inner wall of the column tubing.

- The **mobile phase** is the phase that moves in a definite direction. It may be a liquid (LC and Capillary Electrochromatography (CEC)), a gas (GC), or a supercritical fluid (supercritical-fluid chromatography, SFC). The mobile phase consists of the sample being separated/analyzed and the solvent that moves the sample through the column. In the case of HPLC the mobile phase consists of a non-polar solvent(s) such as hexane in normal phase or a polar solvent such as methanol in reverse phase chromatography and the sample being separated. The mobile phase moves through the chromatography column (the stationary phase) where the sample interacts with the stationary phase and is separated.
- **Preparative chromatography** is used to purify sufficient quantities of a substance for further use, rather than analysis.
- The **retention time** is the characteristic time it takes for a particular analyte to pass through the system (from the column inlet to the detector) under set conditions. See also: Kovats` retention index
- The **sample** is the matter analyzed in chromatography. It may consist of a single component or it may be a mixture of components. When the sample is treated in the course of an analysis, the phase or the phases containing the analytes of interest is/are referred to as the sample whereas everything out of interest separated from the sample before or in the course of the analysis is referred to as waste.
- The **solute** refers to the sample components in partition chromatography.
- The **solvent** refers to any substance capable of solubilizing another substance, and especially the liquid mobile phase in liquid chromatography.
- The **stationary phase** is the substance fixed in place for the chromatography procedure. Examples include the silica layer in thin layer chromatography
- The **detector** refers to the instrument used for qualitative and quantitative detection of analytes after separation.

Chromatography is based on the concept of partition coefficient. Any solute partitions between two immiscible solvents. When we make one solvent immobile (by adsorption on a solid support matrix) and another mobile it results in most common applications of chromatography. If the matrix support, or stationary phase, is polar (e.g. paper, silica etc.) it is forward phase chromatography, and if it is non-polar (C-18) it is reverse phase.

DISPLACEMENT CHROMATOGRAPHY

The basic principle of displacement chromatography is: A molecule with a high affinity for the chromatography matrix (the displacer) competes effectively for binding sites, and thus displace all molecules with lesser affinities. There are distinct differences between displacement and elution chromatography. In elution mode, substances typically emerge from a column in narrow, Gaussian peaks. Wide separation of peaks, preferably to baseline, is desired for maximum purification. The speed at which any component of a mixture travels down the column in elution mode depends on many factors. But for two substances to travel at different speeds, and thereby be resolved, there must be substantial differences in some interaction between the biomolecules and the chromatography matrix. Operating parameters are adjusted to maximize the effect of this difference. In many cases, baseline separation of the peaks can be achieved only with gradient elution and low column loadings. Thus, two drawbacks to elution mode chromatography, especially at the preparative scale, are operational complexity, due to gradient solvent pumping, and low throughput, due to low column loadings. Displacement chromatography has advantages over elution chromatography in that components are resolved into consecutive zones of pure substances rather than "peaks". Because the process takes advantage of the nonlinearity of the isotherms, a larger column feed can be separated on a given column with the purified components recovered at significantly higher concentrations.

TECHNIQUES BY SEPARATION MECHANISM

Ion Exchange Chromatography

Ion exchange chromatography (usually referred to as ion chromatography) uses an ion exchange mechanism to separate analytes based on their respective charges. It is usually performed in columns but can also be useful in planar mode. Ion exchange chromatography uses a charged stationary phase to separate charged compounds including anions, cations, amino acids, peptides, and proteins. In conventional methods the stationary phase is an ion exchange resin that carries charged functional groups that interact with oppositely charged groups of the compound to retain. Ion exchange chromatography is commonly used to purify proteins using FPLC.

Size-Exclusion Chromatography

Size-exclusion chromatography (SEC) is also known as **gel permeation chromatography** (GPC) or **gel filtration chromatography** and separates molecules according to their size (or more accurately according to their hydrodynamic diameter or hydrodynamic volume). Smaller molecules are able to enter the pores of the media and, therefore, molecules are trapped and removed from the flow of the mobile phase. The average residence time in the pores depends upon the effective size of the analyte molecules. However, molecules that are larger than the average pore size of the packing are excluded and thus suffer essentially no retention; such species are the first to be eluted. It is generally a low-resolution chromatography technique and thus it is often reserved for the final,polishing step of a purification. It is also useful for determining the tertiary structure and quaternary structure of purified proteins, especially since it can be carried out under native solution conditions.

Expanded Bed Adsorption Chromatographic Separation

An expanded bed chromatographic adsorption (EBA) column for a biochemical separation process comprises a pressure equalization liquid distributor having a self-cleaning function below a porous blocking sieve plate at the bottom of the expanded bed, an upper part nozzle assembly having a backflush cleaning function at the top of the expanded bed, a better distribution of the feedstock liquor added into the expanded bed ensuring that the fluid passed through the expanded bed layer displays a state of piston flow. The expanded bed layer displays a state of piston flow. The expanded bed chromatographic separation column has advantages of increasing the separation efficiency of the expanded bed.

Expanded-bed adsorption (EBA) chromatography is a convenient and effective technique for the capture of proteins directly from unclarified crude sample. In EBA chromatography, the settled bed is first expanded by upward flow of equilibration buffer. The crude feed, a mixture of soluble proteins, contaminants, cells, and cell debris, is then passed upward through the expanded bed. Target proteins are captured on the adsorbent, while particulates and contaminants pass through. A change to elution buffer while maintaining upward flow results in desorption of the target protein in expanded-bed mode. Alternatively, if the flow is reversed, the adsorbed particles will quickly settle and the proteins can be desorbed by an elution buffer. The mode used for elution (expanded-bed versus settled-bed) depends on the characteristics of the feed. After elution, the adsorbent is cleaned with a predefined cleaning-in-place (CIP) solution, with cleaning followed by either column regeneration (for further use) or storage.

REFERENCES

1. chromatography. Online Etymology Dictionary.
2. McMurry, John (2011). Organic chemistry: with biological applications (2nd ed.). Belmont, CA: Brooks/Cole. p. 395. ISBN 9780495391470.
3. Hostettmann, K; Marston, A; Hostettmann, M (1998). Preparative Chromatography Techniques Applications in Natural Product Isolation (Second ed.). Berlin, Heidelberg: Springer Berlin Heidelberg. p. 50. ISBN 9783662036310.
4. The first report of Tsvet`s discovery was published in 1905:
5. Ettre, L. S.; Sakodynskii, K. I. (March 1993).M. S. Tswett and the discovery of chromatography II: Completion of the development of chromatography (1903–1910). Chromatographia. 35 *(5-6): 329–338.* doi:10.1007/BF02277520.
6. The Nobel Prize in Chemistry 1952. nobelprize.org. Retrieved 25 August 2016.
7. Ettre, L. S. (1993).Nomenclature for chromatography (IUPAC Recommendations 1993). Pure and Applied Chemistry. 65 *(4).* doi:10.1351/pac199365040819.
8. Still, W. C.; Kahn, M.; Mitra, A. (1978).Rapid chromatographic technique for preparative separations with moderate resolution. J. Org. Chem. 43 *(14): 2923–2925.* doi:10.1021/jo00408a041.
9. Harwood, Laurence M.; Moody, Christopher J. (1989). Experimental organic chemistry: Principles and Practice (Illustrated ed.). WileyBlackwell. pp. 180–185. ISBN 978-0-632-02017-1.
10. Anfinsen, Christian B.; Edsall, John Tileston; Richards, Frederic Middlebrook, eds. (1976). Advances in Protein Chemistry. pp. 6–7. ISBN 978-0-12-034230-3.
11. Displacement Chromatography 101 Archived 15 September 2008 at the Wayback Machine.. Sachem, Inc. Austin, TX 78737

12. Bailon, Pascal; Ehrlich, George K.; Fung, Wen-Jian and Berthold, Wolfgang (2000) An Overview of Affinity Chromatography, Humana Press. ISBN 978-0-89603-694-9.

13. Ninfa, Alexander J., David P. Ballou, and Marilee Benore. Fundamental Laboratory Approaches for Biochemistry and Biotechnology. Hoboken, NJ: John Wiley, 2010. Print.

14. Müller, Tobias K.h., and Matthias Franzreb.Suitability of Commercial Hydrophobic Interaction Sorbents for Temperature-controlled Protein Liquid Chromatography under Low Salt Conditions. Journal of Chromatography A 1260 (2012): 88-96. Web. <http://ac.els-cdn.com/S002196731201268X/1-s2.0-S002196731201268X-main.pdf?_tid=2f7e7186-1c7e-11e6-836f-00000aacb361&acdnat=1463524077_be16e37f7efe44d6b689342fa7669726>.

15. Ren, Jun, Peng Yao, Jingjing Chen, and Lingyun Jia.Salt-independent Hydrophobic Displacement Chromatography for Antibody Purification Using Cyclodextrin as Supermolecular Displacer. Journal of Chromatography A 1369 (2014): 98-104. Web. <http://ac.els-cdn.com/S0021967314015684/1-s2.0-S0021967314015684-main.pdf?_tid=b4e118f2-1c87-11e6-ba07-00000aab0f27&acdnat=1463528166_e445fcef926f1c22934169b8818c53c6>.

CHAPTER 2

Distribution Constant

Chapter Contents

- **Introduction**
- **Freundlich Equation**
- **Kovats Retention Index**
- **Retardation Factor**
- **Van Deemter Equation**

The distribution constant is the concentration of a component in or on the stationary phase divided by the concentration of the component in the mobile phase in equilibrium conditions. Since in chromatography a component may be present in more than one form (e.g., associated and dissociated forms), the analytical condition used here refers to the total amount present without regard to the existence of various forms. These terms are also called the Distribution Coefficients. However, the present term conforms more closely to the general usage in science.

INTRODUCTION

The **distribution constant** (or **partition ratio**) (**K_D**), is the equilibrium constant for the distribution of an analyte in two immiscible solvents.

In chromatography, for a particular solvent, it is equal to the ratio of its molar concentration in the stationary phase to its molar concentration in the mobile phase, also approximating the ratio of the solubility of the solvent in each phase.

The term is often confused with partition coefficient or distribution coefficient.

Expression

The ratio of activities of a solute, A in an aqueous/organic system will remain constant and independent of the total quantity of A (hence $[A]_{org} \propto [A]_{aq}$), so at any given temperature:

$$(K_D)_A = \frac{(aA)_{org}}{(a_A)_{aq}} \approx \frac{[A]_{org}}{[A]_{aq}}$$

Distribution constants are useful as they allow the calculation of the concentration of remaining analyte in the solution, even after a number of solvent extractions have occurred. They also provide guidance in choosing the most efficient way to conduct an extractive separation.

Thus, the concentration of A remaining in an aqueous solution after *i* extractions with an organic solvent can be found using:

$$[A]_i = (\frac{V_{aq}}{V_{org}K_D + V_{aq}})^i [A]_0$$

(where $[A]_i$ is the concentration of A remaining after extracting V_{aq} millilitres of solution with the original concentration of $[A]_0$ with *i* portions of the organic solvent, each with a volume of V_{org}).

FREUNDLICH EQUATION

The **Freundlich equation** or **Freundlich adsorption isotherm**, an adsorption isotherm, is an empirical relation between the concentration of a solute on the surface of an adsorbent to the concentration of the solute in the liquid with which it is in contact. In 1909, Herbert Freundlich gave an expression representing the isothermal variation of adsorption of a quantity of gas adsorbed by unit mass of solid adsorbent with pressure. This equation is known as Freundlich adsorption isotherm or Freundlich adsorption equation. As this relationship is entirely empirical, in the case where adsorption behavior can be properly fit by isotherms with a theoretical basis, it is usually appropriate to use such isotherms instead (see for example the Langmuir and BET adsorption theories).

Freundlich Adsorption Isotherm

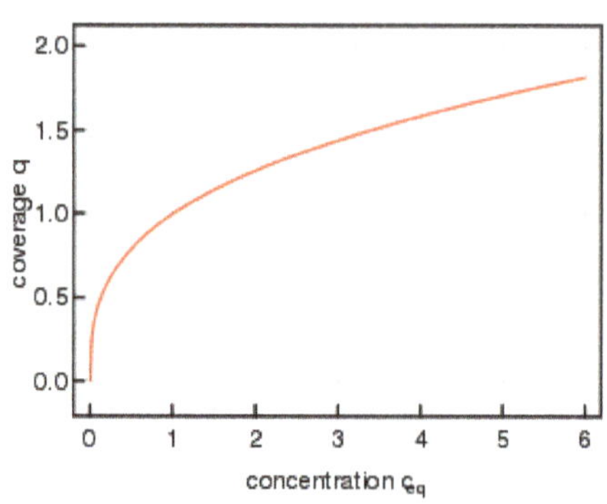

An isotherm according to the Freundlich equation with concstants K=1 and n=3.

The Freundlich adsorption isotherm is mathematically expressed as

$$\frac{x}{m} = Kp^{1/n}$$

It is also written as

$$\log \frac{x}{m} = \log K + \frac{1}{n} \log p$$

or

$$\frac{x}{m} = Kc^{1/n}$$

It is also written as

$$\log \frac{x}{m} = \log K + \frac{1}{n} \log c$$

where

x = mass of adsorbate

m = mass of adsorbent

p = Equilibrium pressure of adsorbate

c = Equilibrium concentration of adsorbate in solution.

K and n are constants for a given adsorbate and adsorbent at a particular temperature.

At high pressure 1/n = 0, hence extent of adsorption becomes independent of pressure. It is used in cases where the actual identity of the solute is not known, such as adsorption of colored material from sugar, vegetable oil etc.

Limitation of Freundlich Adsorption Isotherm

Experimentally it was determined that extent of adsorption varies directly with pressure and then it directly varies with pressure raised to the power 1/n until saturation pressure P_s is reached. Beyond that point rate of adsorption saturates even after applying higher pressure. Thus Freundlich adsorption isotherm failed at higher pressure.

KOVATS RETENTION INDEX

In gas chromatography, **Kovats retention index** (shorter **Kovats index**, **retention index**; plural **retention indices**) is used to convert retention times into system-independent constants. The index is named after the Hungarian-born Swiss chemist Ervin Kováts, who outlined this concept during the 1950s while performing research into the composition of the essential oils.

The retention index of a certain chemical compound is its retention time normalised to the retention times of adjacently eluting n-alkanes. While retention times vary with the individual chromatographic system (e.g. with regards to column length, film thickness, diameter, carrier gas velocity and pressure, and void time), the derived retention indices are quite independent of these parameters and allow comparing values measured by different analytical laboratories under varying conditions. Tables of retention indices can help identify components by comparing experimentally found retention indices with known values.

Expression

The method takes advantage of the linear relationship between the values of $\log(t'_r)$ and the number of carbon atoms in a molecule. The value of Kovats index is usually represented by I in mathematical expressions. Its applicability is restricted to organic compounds. For isothermal chromatography, the Kovats index is given by the equation

$$I = 100 \times \left[n + \frac{\log(t'_{r(\text{unknown})}) - \log(t'_{r(n)})}{\log(t'_{r(N)}) - \log(t'_{r(n)})} \right]$$

Where:

I Kovats retention index,

n the number of carbon atoms in the smaller n-alkane,

N the number of carbon atoms in the larger n-alkane,

t'_r the adjusted retention time.

For temperature programmed chromatography, the Kovats index is given by the equation

$$I = 100 \times \left[n + \frac{t_{r(\text{unknown})} - t_{r(n)}}{t_{r(N)} - t_{r(n)}} \right]$$

Where:

I Kovats retention index,

n the number of carbon atoms in the smaller n-alkane,

N the number of carbon atoms in the larger n-alkane,

t'_r the retention time.

RETARDATION FACTOR

In chromatography, the **retardation factor** (***R***) is the fraction of an analyte in the mobile phase of a chromatographic system. In planar chromatography in particular, the retardation factor R_f is defined as the ratio of the distance traveled by the center of a spot to the distance traveled by the solvent front. Ideally, the values for R_F are equivalent to the R values used in column chromatography.

Although the term **retention factor** is sometimes used synonymously with retardation factor in regard to planar chromatography the term is not defined in this context. However, in column chromatography, the retention factor or **capacity factor** (***k***) is defined as the ratio of time an analyte is retained in the stationary phase to the time it is retained in the mobile phase, which is inversely proportional to the retardation factor.

General Definition

In chromatography the retardation factor, R, is the fraction of the sample in the mobile phase at equilibrium, defined as:

$$R = \frac{\text{quantity of substance in the mobile phase}}{\text{total quantity of substance in the system}}$$

Planar Chromatography

The retardation factor, R_f, is commonly used in paper chromatography and thin layer chromatography for analyzing and comparing different substances. It can can be mathematically described by the following ratio:

$$R_f = \frac{\text{migration distance of substance}}{\text{migration distance of solvent front}}$$

An R_f value will always be in the range 0 to 1; if the substance moves, it can only move in the direction of the solvent flow, and cannot move faster than the solvent. For example, if particular substance in an unknown mixture travels 2.5 cm and the solvent front travels 5.0 cm, the retardation factor would be 0.5. One can choose a mobile phase with different characteristics (particularly polarity) in order to control how far the substance being investigated migrates.

An R_f value is characteristic for any given compound (provided that the same stationary and mobile phases are used). It can provide corroborative evidence as to the identity of a compound. If the identity of a compound is suspected but not yet proven, an authentic sample of the compound, or standard, is spotted and run on a TLC plate side by side (or on top of each other) with the compound in question. Note that this identity check must be performed on a single plate, because it is difficult to duplicate all the factors which influence R_f exactly from experiment to experiment.

Relationship with retention factor

In terms of retention factor (k), retardation factor (R) is defined as follows:

$$R = \frac{1}{k+1}$$

and conversely:

$$k = \frac{1-R}{R}$$

VAN DEEMTER EQUATION

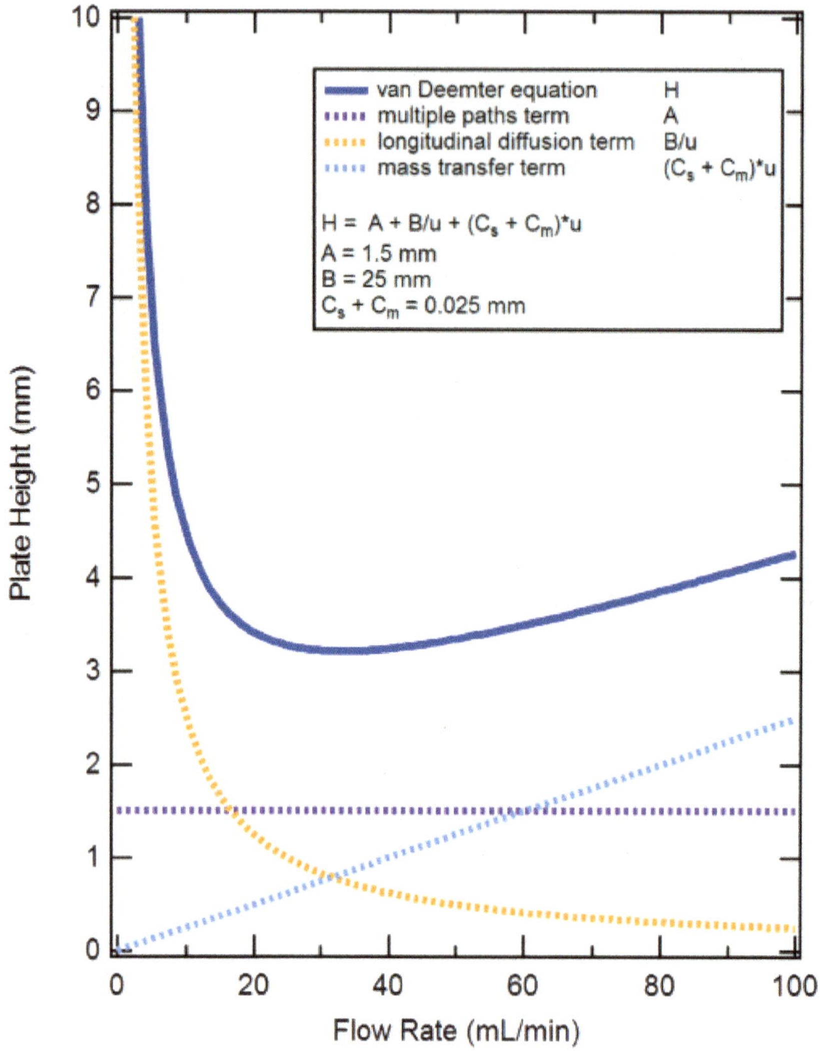

The **Van Deemter equation** in chromatography relates the variance per unit length of a separation column to the linear mobile phase velocity by considering physical, kinetic, and thermodynamic properties of a separation. These properties include pathways within the column, diffusion (axial and longitudinal), and mass transfer kinetics between stationary and mobile phases. In liquid chromatography, the mobile phase velocity is taken as the exit velocity, that is, the ratio of the flow rate in ml/second to the cross-sectional area of the 'column-exit flow path.' For a packed column, the cross-sectional area of the column exit flow path is usually taken as 0.6 times the cross-sectional area of the column. Alternatively, the linear velocity can be taken as the ratio of the column length to the dead time. If the mobile phase is a gas, then the pressure correction must be applied. The variance per unit length of the column is taken as the ratio of the column length to the column efficiency in theoretical plates. The Van Deemter equation is a hyperbolic function that predicts that there is an optimum velocity at which there will be the minimum variance per unit column length and, thence, a maximum efficiency. The Van Deemter equation was the result of the first application of rate theory to the chromatography elution process.

Van Deemter equation

The van Deemter equation relates the resolving power (HETP, height equivalent to a theoretical plate) of a chromatographic column to the various flow and kinetic parameters which cause peak broadening, as follows:

$$HETP = A + \frac{B}{u} + (C_s + C_m) \cdot u$$

Where

- HETP = height equivalent to a theoretical plate, a measure of the resolving power of the column [m]
- A = Eddy-diffusion parameter, related to channeling through a non-ideal packing [m]
- B = diffusion coefficient of the eluting particles in the longitudinal direction, resulting in dispersion [$m^2 s^{-1}$]
- C = Resistance to mass transfer coefficient of the analyte between mobile and stationary phase [s]
- u = Linear Velocity [$m\ s^{-1}$]

In open tubular capillaries, the A term will be zero as the lack of packing means channeling does not occur. In packed columns, however, multiple distinct routes (channels) exist through the column packing, which results in band spreading. In the latter case, A will not be zero.

The form of the van Deemter equation is such that HETP achieves a minimum value at a particular flow velocity. At this flow rate, the resolving power of the column is maximized, although in practice, the elution time is likely to be impractical. Differentiating the van Deemter equation with respect to velocity, setting the resulting expression equal to zero, and solving for the optimum velocity yields the following:

$$u = \sqrt{\frac{B}{C}}$$

Plate Count

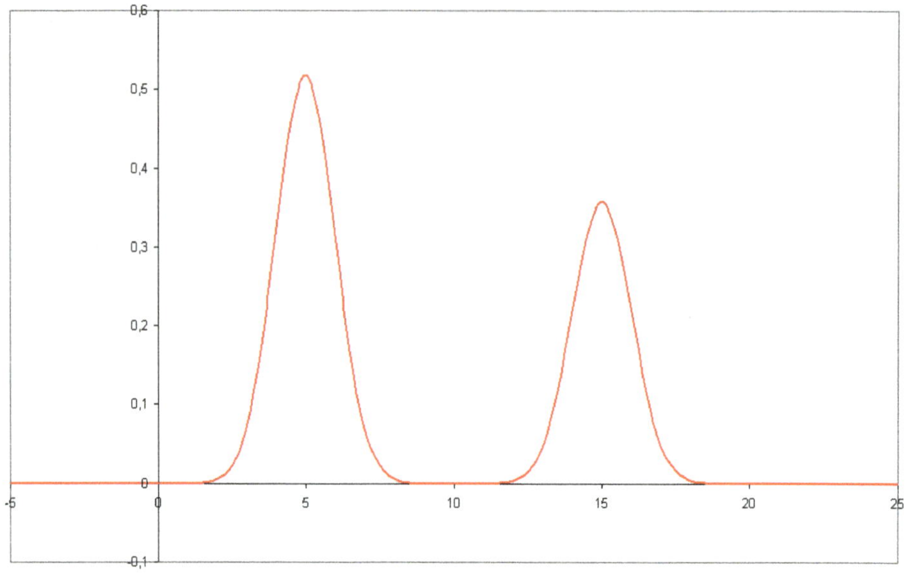

Two well resolved peaks in a chromatogram.

The plate height given as:

$$H = \frac{L}{N}$$

with L the column length and N the number of theoretical plates can be estimated from a chromatogram by analysis of the retention time t_R for each component and its standard deviation σ as a measure for peak width, provided that the elution curve represents a Gaussian curve.

In this case the plate count is given by:

$$N = \left(\frac{t_R}{\sigma}\right)^2$$

By using the more practical peak width at half height $W_{1/2}$ the equation is:

$$N = 8\ln(2) \cdot \left(\frac{t_R}{W_{1/2}}\right)^2$$

or with the width at the base of the peak:

$$N = 16 \cdot \left(\frac{t_R}{W_{base}}\right)^2$$

Expanded van Deemter

The Van Deemter equation can be further expanded to:

$$H = 2\lambda d_p + \frac{2\gamma D_m}{u} + \frac{\omega(d_p \text{ or } d_c)^2 u}{D_m} + \frac{R d_f^2 u}{D_s}$$

Where:

- H is plate height
- λ is particle shape (with regard to the packing)
- d_p is particle diameter
- γ, ω, and R are constants
- D_m is the diffusion coefficient of the mobile phase
- d_c is the capillary diameter
- d_f is the film thickness
- D_s is the diffusion coefficient of the stationary phase.
- u is the linear velocity

REFERENCES

1. Institute of Chemical Sciences and Engineering, KOVATS Ervin, Prof. Hon.
2. *Kovats, E. (1958). Gas-chromatographische Charakterisierung organischer Verbindungen. Teil 1: Retentionsindices aliphatischer Halogenide, Alkohole, Aldehyde und Ketone. Helv. Chim. Acta.* ***41*** *(7): 1915–32. doi:10.1002/hlca.19580410703.*
3. IUPAC, *Compendium of Chemical Terminology*, 2nd ed. (the "Gold Book") (1997). Online corrected version: (2006–) "retention index, **I** in column chromatography".
4. Retention index guide

CHAPTER 3

Techniques by Chromatographic Bed Shape

Chapter Contents

- **Introduction**
- **High-Performance Liquid Chromatography**
- **Operation**
- **Isocratic and Gradient Elution**
- **Parameters**
- **Applications**
- **Fast Protein Liquid Chromatography**

Chromatography may be preparative or analytical. The purpose of preparative chromatography is to separate the components of a mixture for more advanced use (and is thus a form of purification). Analytical chromatography is done normally with smaller amounts of material and is for measuring the relative proportions of analytes in a mixture. The two are not mutually exclusive. There are different kinds of chromatographic techniques and these are classified according to the shape of bed, physical state of mobile phase, separation mechanisms.

Apart from these there are certain modified forms of these chromatographic techniques involving different mechanisms and are hence categorized as modified or specialized chromatographic techniques.

INTRODUCTION

Column chromatography is a separation technique in which the stationary bed is within a tube. The particles of the solid stationary phase or the support coated with a liquid stationary phase may fill the whole inside volume of the tube (packed column) or be concentrated on or along the inside tube wall leaving an open, unrestricted path for the mobile phase in the middle part of the tube (open tubular column). Differences in rates of movement through the medium are calculated to different retention times of the sample.

In 1978, W. Clark Still introduced a modified version of column chromatography called **flash column chromatography** (flash). The technique is very similar to the traditional column chromatography, except for that the solvent is driven through the column by applying positive pressure. This allowed most separations to be performed in less than 20 minutes, with improved separations compared to the old method. Modern flash chromatography systems are sold as pre-packed plastic cartridges, and the solvent is pumped through the cartridge. Systems may also be linked with detectors and fraction collectors providing automation. The introduction of gradient pumps resulted in quicker separations and less solvent usage.

In expanded bed adsorption, a fluidized bed is used, rather than a solid phase made by a packed bed. This allows omission of initial clearing steps such as centrifugation and filtration, for culture broths or slurries of broken cells.

Phosphocellulose chromatography utilizes the binding affinity of many DNA-binding proteins for phosphocellulose. The stronger a protein`s interaction with DNA, the higher the salt concentration needed to elute that protein.

Column Chromatography

Column chromatography in chemistry is a method used to purify individual chemical compounds from mixtures of compounds. It is often used for preparative applications on scales from micrograms up to kilograms. The main advantage of column chromatography is the relatively low cost and disposability of the stationary phase used in the process. The latter prevents cross-contamination and stationary phase degradation due to recycling.

The classical preparative chromatography column is a glass tube with a diameter from 5 mm to 50 mm and a height of 5 cm to 1 m with a tap and some kind of a filter (a glass frit or glass wool plug – to prevent the loss of the stationary phase) at the bottom. Two methods are generally used to prepare a column: the dry method and the wet method.

- For the dry method, the column is first filled with dry stationary phase powder, followed by the addition of mobile phase, which is flushed through the column until it is completely wet, and from this point is never allowed to run dry.
- For the wet method, a slurry is prepared of the eluent with the stationary phase powder and then carefully poured into the column. Care must be taken to avoid air bubbles. A solution of the organic material is pipetted on top of the stationary phase. This layer is usually topped with a small layer of sand or with cotton or glass wool to protect the shape of the organic layer from the velocity of newly added eluent. Eluent is slowly passed through the column to advance the organic material. Often a spherical eluent reservoir or an eluent-filled and stoppered separating funnel is put on top of the column.

A chemist in the 1950s using column chromatography. The Erlenmeyer receptacles are on the floor.

The individual components are retained by the stationary phase differently and separate from each other while they are running at different speeds through the column with the eluent. At the end of the column they elute one at a time. During the entire chromatography process the eluent is collected in a series of fractions. Fractions can be collected automatically by means of fraction collectors. The productivity of chromatography can be increased by running several columns at a time. In this case multi stream collectors are used. The composition of the eluent flow can be monitored and each fraction is analyzed for dissolved compounds, e.g. by analytical chromatography, UV absorption spectra, or fluorescence. Colored compounds (or fluorescent compounds with the aid of a UV lamp) can be seen through the glass wall as moving bands.

Stationary Phase

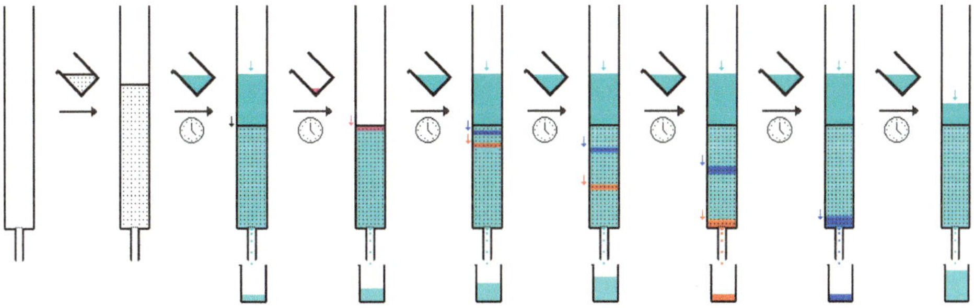

Column chromatography proceeds by a series of steps.

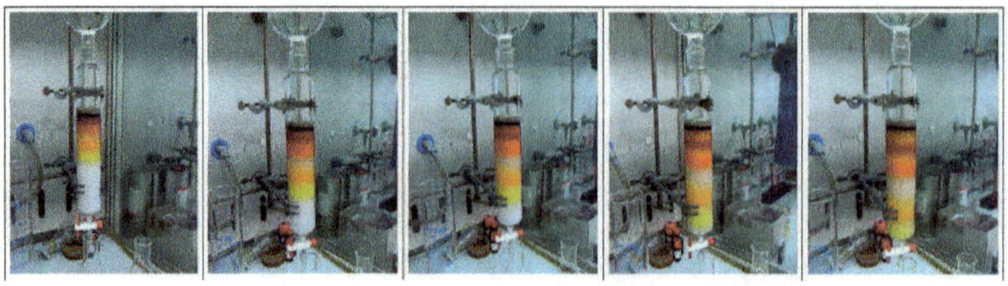

Photographic sequence of a column chromatography.

The *stationary phase* or *adsorbent* in column chromatography is a solid. The most common stationary phase for column chromatography is silica gel, the next most common being alumina. Cellulose powder has often been used in the past. Also possible are ion exchange chromatography, reversed-phase chromatography (RP), affinity chromatography or expanded bed adsorption (EBA). The stationary phases are usually finely ground powders or gels and/or are microporous for an increased surface, though in EBA a fluidized bed is used. There is an important ratio between the stationary phase weight and the dry weight of the analyte mixture that can be applied onto the column. For silica column chromatography, this ratio lies within 20:1 to 100:1, depending on how close to each other the analyte components are being eluted.

Mobile Phase (Eluent)

The *mobile phase* or *eluent* is either a pure solvent or a mixture of different solvents. It is chosen so that the retention factor value of the compound of interest is roughly around 0.2 - 0.3 in order to minimize the time and the amount of eluent to run the chromatography. The eluent has also been chosen so that the different compounds can be separated effectively. The eluent is optimized in small scale pretests, often using thin layer chromatography (TLC) with the same stationary phase.

There is an optimum flow rate for each particular separation. A faster flow rate of the eluent minimizes the time required to run a column and thereby minimizes diffusion, resulting in a better separation. However, the maximum flow rate is limited because a finite time is required for the analyte to equilibrate between the stationary phase and mobile phase, see Van Deemter`s equation. A simple laboratory column runs by gravity flow. The flow rate of such a column can be increased by extending the fresh eluent filled column above the top of the stationary phase or decreased by the tap controls. Faster flow rates can be achieved by using a pump or by using compressed gas (e.g. air, nitrogen, or argon) to push the solvent through the column (flash column chromatography).

The particle size of the stationary phase is generally finer in flash column chromatography than in gravity column chromatography. For example, one of the most widely used silica gel grades in the former technique is mesh 230 – 400 (40 – 63 μm), while the latter technique typically requires mesh 70 – 230 (63 – 200 μm) silica gel.

A spreadsheet that assists in the successful development of flash columns has been developed. The spreadsheet estimates the retention volume and band volume of analytes, the fraction numbers expected to contain each analyte, and the resolution between adjacent peaks. This information allows users to select optimal parameters for preparative-scale separations before the flash column itself is attempted.

Automated Systems

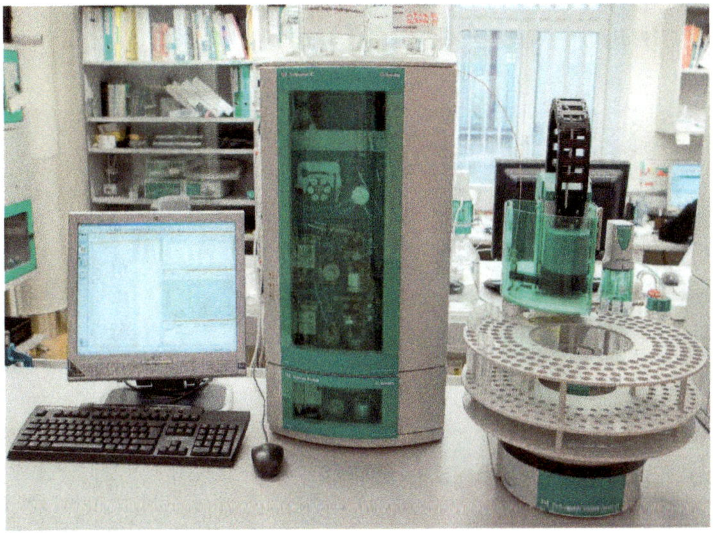

An automated ion chromatography system.

Column chromatography is an extremely time consuming stage in any lab and can quickly become the bottleneck for any process lab. Therefore, several manufacturers like Biotage, Buchi, Interchim and Teledyne Isco have developed automated flash chromatography systems (typically referred to as LPLC, low pressure liquid chromatography, around 350–525 kPa or 50.8–76.1 psi) that minimize human involvement in the purification process. Automated systems will include components normally found on more expensive high performance liquid chromatography (HPLC) systems such as a gradient pump, sample injection ports, a UV detector and a fraction collector to collect the eluent. Typically these automated systems can separate samples from a few milligrams up to an industrial many kilogram scale and offer a much cheaper and quicker solution to doing multiple injections on prep-HPLC systems.

The resolution (or the ability to separate a mixture) on an LPLC system will always be lower compared to HPLC, as the packing material in an HPLC column can be much smaller, typically only 5 micrometre thus increasing stationary phase surface area, increasing surface interactions and giving better separation. However, the use of this small packing media causes the high back pressure and is why it is termed high pressure liquid chromatography. The LPLC columns are typically packed with silica of around 50 micrometres, thus reducing back pressure and resolution, but it also removes the need for expensive high pressure pumps. Manufacturers are now starting to move into higher pressure flash chromatography systems and have termed these as medium pressure liquid chromatography (MPLC) systems which operate above 1 MPa (150 psi).

Column Chromatogram Resolution Calculation

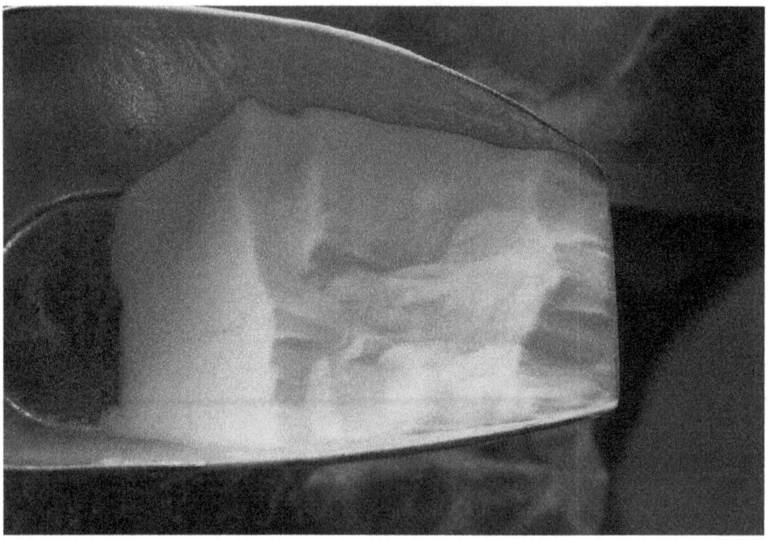

Powdery silica gel for column chromatography.

Typically, column chromatography is set up with peristaltic pumps, flowing buffers and the solution sample through the top of the column. The solutions and buffers pass through the column where a fraction collector at the end of the column setup collects the eluted samples. Prior to the fraction collection, the samples that are eluted from the column pass through a detector such as a spectrophotometer or mass spectrometer so that the concentration of the separated samples in the sample solution mixture can be determined.

For example, if you were to separate two different proteins with different binding capacities to the column from a solution sample, a good type of detector would be a spectrophotometer using a wavelength of 280 nm. The higher the concentration of protein that passes through the eluted solution through the column, the higher the absorbance of that wavelength.

Because the column chromatography has a constant flow of eluted solution passing through the detector at varying concentrations, the detector must plot the concentration of the eluted sample over a course of time. This plot of sample concentration versus time is called a chromatogram.

The ultimate goal of chromatography is to separate different components from a solution mixture. The resolution expresses the extent of separation between the components from the mixture. The higher the resolution of the chromatogram, the better the extent of separation of the samples the column gives. This data is a good way of determining the column's separation properties of that particular sample. The resolution can be calculated from the chromatogram.

The separate curves in the diagram represent different sample elution concentration profiles over time based on their affinity to the column resin. To calculate resolution, the retention time and curve width are required.

Retention time: The time from the start of signal detection by the detector to the peak height of the elution concentration profile of each different sample.

Curve width: The width of the concentration profile curve of the different samples in the chromatogram in units of time.

A simplified method of calculating chromatogram resolution is to use the plate model. The plate model assumes that the column can be divided into a certain number of sections, or plates and the mass balance can be calculated for each individual plate. This approach approximates a typical chromatogram curve as a Gaussian distribution curve. By doing this, the curve width is estimated as 4 times the standard deviation of the curve, 4σ. The retention time is the time from the start of signal detection to the time of the peak height of the Gaussian curve.

From the variables in the figure above, the resolution, plate number, and plate height of the column plate model can be calculated using the equations:

Resolution (R_s)

$R_s = 2(t_{RB} - t_{RA})/(w_B + w_A)$

Where:

t_{RB} = retention time of solute B

t_{RA} = retention time of solute A

w_B = Gaussian curve width of solute B

w_A = Gaussian curve width of solute A

Plate Number (N):

$N = (t_R)^2/(w/4)^2$

Plate Height (H):

$H = L/N$

Where L is the length of the column.

Column Adsorption Equilibrium

For an adsorption column, the column resin (the stationary phase) is composed of microbeads. Even smaller particles such as proteins, carbohydrates, metal ions, or other chemical compounds are conjugated onto the microbeads. Each binding particle that is attached to the microbead can be assumed to bind in a 1:1 ratio with the solute sample sent through the column that needs to be purified or separated.

Binding between the target molecule to be separated and the binding molecule on the column beads can be modeled using a simple equilibrium reaction $K_{eq} = [CS]/([C][S])$ where K_{eq} is the equilibrium constant, [C] and [S] are the concentrations of the target molecule and the binding molecule on the column resin, respectively. [CS] is the concentration of the complex of the target molecule bound to the column resin.

Using this as a basis, three different isotherms can be used to describe the binding dynamics of a column chromatography: linear, Langmuir, and Freundlich.

The linear isotherm occurs when the solute concentration needed to be purified is very small relative to the binding molecule. Thus, the equilibrium can be defined as:

$[CS] = K_{eq}[C]$.

For industrial scale uses, the total binding molecules on the column resin beads must be factored in because unoccupied sites must be taken into account. The Langmuir isotherm and Freundlich isotherm are useful in describing this equilibrium. Langmuir Isotherm:

$[CS] = (K_{eq}S_{tot}[C])/(1 + K_{eq}[C])$, where S_{tot} is the total binding molecules on the beads.

Freundlich Isotherm:

$[CS] = K_{eq}[C]^{1/n}$

The Freundlich isotherm is used when the column can bind to many different samples in the solution that needs to be purified. Because the many different samples have different binding constants to the beads, there are many different Keq's. Therefore, the Langmuir isotherm is not a good model for binding in this case.

Planar Chromatography

Planar chromatography is a separation technique in which the stationary phase is present as or on a plane. The plane can be a paper, serving as such or impregnated by a substance as the stationary bed (paper chromatography) or a layer of solid particles spread on a support such as a glass plate (thin layer chromatography). Different compounds in the sample mixture travel different distances according to how strongly they interact with the stationary phase as compared to the mobile phase. The specific Retention factor (R_f) of each chemical can be used to aid in the identification of an unknown substance.

Paper Chromatography

Paper chromatography is a technique that involves placing a small dot or line of sample solution onto a strip of *chromatography paper*. The paper is placed in a container with a shallow layer of solvent and sealed. As the solvent rises through the paper, it meets the sample mixture, which starts to travel up the paper with the solvent. This paper is made of cellulose, a polar substance, and the compounds within the mixture travel farther if they are non-polar. More polar substances bond with the cellulose paper more quickly, and therefore do not travel as far.

Thin layer Chromatography (TLC)

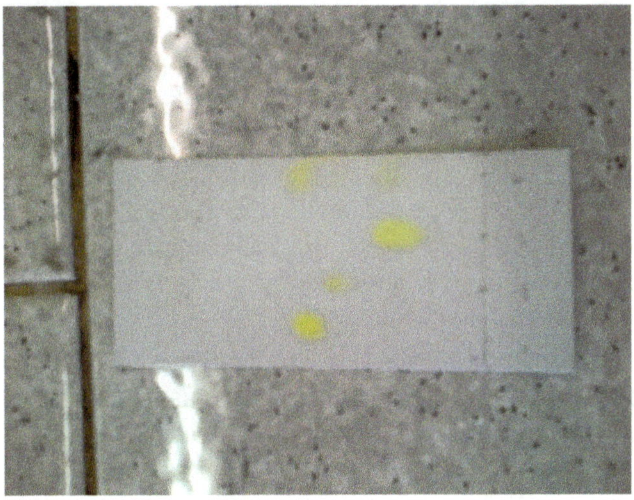

TLC of three standards (ortho-, meta- and para-isomers) and a sample.

fluorescent tlc plate under UV light.

Thin-layer chromatography (TLC) is a chromatography technique used to separate non-volatile mixtures. Thin-layer chromatography is performed on a sheet of glass, plastic, or aluminium foil, which is coated with a thin layer of adsorbent material, usually silica gel, aluminium oxide (alumina), or cellulose. This layer of adsorbent is known as the stationary phase.

After the sample has been applied on the plate, a solvent or solvent mixture (known as the mobile phase) is drawn up the plate via capillary action. Because different analytes ascend the TLC plate at different rates, separation is achieved. The mobile phase has different properties from the stationary phase. For example, with silica gel, a very polar substance, non-polar mobile phases such as heptane are used. The mobile phase may be a mixture, allowing chemists to fine-tune the bulk properties of the mobile phase.

After the experiment, the spots are visualized. Often this can be done simply by projecting ultraviolet light onto the sheet; the sheets are treated with a phosphor, and dark spots appear on the sheet where compounds absorb the light impinging on a certain area. Chemical processes can also be used to visualize spots; anisaldehyde, for example, forms colored adducts with many compounds, and sulfuric acid will char most organic compounds, leaving a dark spot on the sheet.

To quantify the results, the distance traveled by the substance being considered is divided by the total distance traveled by the mobile phase. (The mobile phase must not be allowed to reach the end of the stationary phase.) This ratio is called the **retention factor** or R_f. In general,a substance whose structure resembles the stationary phase will have low R_f, while one that has a similar structure to the mobile phase will have high retention factor. Retention factors are characteristic, but will change depending on the exact condition of the mobile and stationary phase. For this reason, chemists usually apply a sample of a known compound to the sheet before running the experiment.

Thin-layer chromatography can be used to monitor the progress of a reaction, identify compounds present in a given mixture, and determine the purity of a substance. Specific examples of these applications include: analyzing ceramides and fatty acids, detection of pesticides or insecticides in food and water, analyzing the dye composition of fibers in forensics, assaying the radiochemical purity of radiopharmaceuticals, or identification of medicinal plants and their constituents A number of enhancements can be made to the original method to automate the different steps, to increase the resolution achieved with TLC and to allow more accurate quantitative analysis. This method is referred to as HPTLC, or high-performance TLC". HPTLC typically uses thinner layers of stationary phase and smaller sample volumes, thus reducing the loss of resolution due to diffusion.

Plate Preparation

TLC plates are usually commercially available, with standard particle size ranges to improve reproducibility. They are prepared by mixing the adsorbent, such as silica gel, with a small amount of inert binder like calcium sulfate (gypsum) and water. This mixture is spread as a thick slurry on an unreactive carrier sheet, usually glass, thick aluminum foil, or plastic. The resultant plate is dried and *activated* by heating in an oven for thirty minutes at 110 °C. The thickness of the absorbent layer is typically around 0.1 – 0.25 mm for analytical purposes and around 0.5 – 2.0 mm for preparative TLC.

Technique

The process is similar to paper chromatography with the advantage of faster runs, better separations, and the choice between different stationary phases. Because of its simplicity and speed TLC is often used for monitoring chemical reactions and for the qualitative analysis of reaction products. Plates can be labeled before or after the chromatography process using a pencil or other implement that will not interfere or react with the process.

To run a thin layer chromatography plate, the following procedure is carried out:

- Using a capillary, a small spot of solution containing the sample is applied to a plate, about 1.5 centimeters from the bottom edge. The solvent is allowed to completely evaporate off to prevent it from interfering with sample's interactions with the mobile phase in the next step. If a non-volatile solvent was used to apply the sample, the plate needs to be dried in a vacuum chamber. This step is often repeated to ensure there is enough analyte at the starting spot on the plate to obtain a visible result. Different samples can be placed in a row of spots the same distance from the bottom edge, each of which will move in its own adjacent lane from its own starting point.
- A small amount of an appropriate solvent (eluent) is poured into a glass beaker or any other suitable transparent container (separation chamber) to a depth of less than 1 centimeter. A strip of filter paper (aka wick") is put into the chamber so that its bottom touches the solvent and the paper lies on the chamber wall and reaches almost to the top of the container. The container is closed with a cover glass or any other lid and is left for a few minutes to let the solvent vapors ascend the filter paper and saturate the air in the chamber. (Failure to saturate the chamber will result in poor separation and non-reproducible results).
- The TLC plate is then placed in the chamber so that the spot(s) of the sample do not touch the surface of the eluent in the chamber, and the lid is closed. The solvent moves up the plate by capillary action, meets the sample mixture and carries it up the plate (elutes the sample). The plate should be removed from the chamber before the solvent front reaches the top of the stationary phase (continuation of the elution will give a misleading result) and dried.
- Without delay, the *solvent front*, the furthest extent of solvent up the plate, is marked.
- The plate is visualized. As some plates are pre-coated with a phosphor such as zinc sulfide, allowing many compounds to be visualized by using ultraviolet light; dark spots appear where the compounds block the UV light from striking the plate. Alternatively, plates can be sprayed or immersed in chemicals after elution. Various visualising agents react with the spots to produce visible results.

Separation Process and Principle

Different compounds in the sample mixture travel at different rates due to the differences in their attraction to the stationary phase and because of differences in solubility in the solvent. By changing the solvent, or perhaps using a mixture, the separation of components (measured by the R_f value) can be adjusted. Also, the separation achieved with a TLC plate can be used to estimate the separation of a flash chromatography column. (A compound elutes from a column when the amount of solvent collected is equal to $1/R_f$.) Chemists often use TLC to develop a protocol for separation by chromatography and they use TLC to determine which fractions contain the desired compounds.

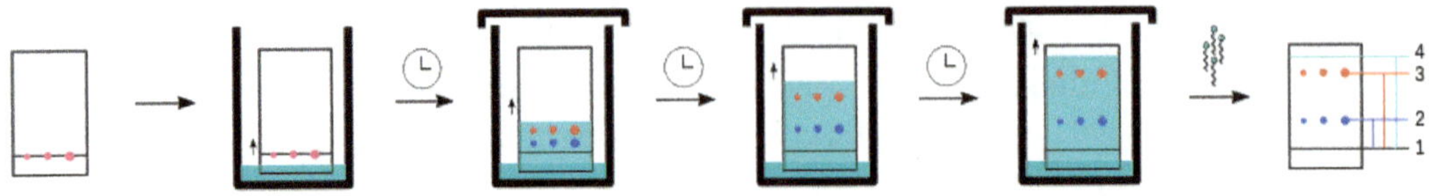

Development of a TLC plate, a purple spot separates into a red and blue spot.

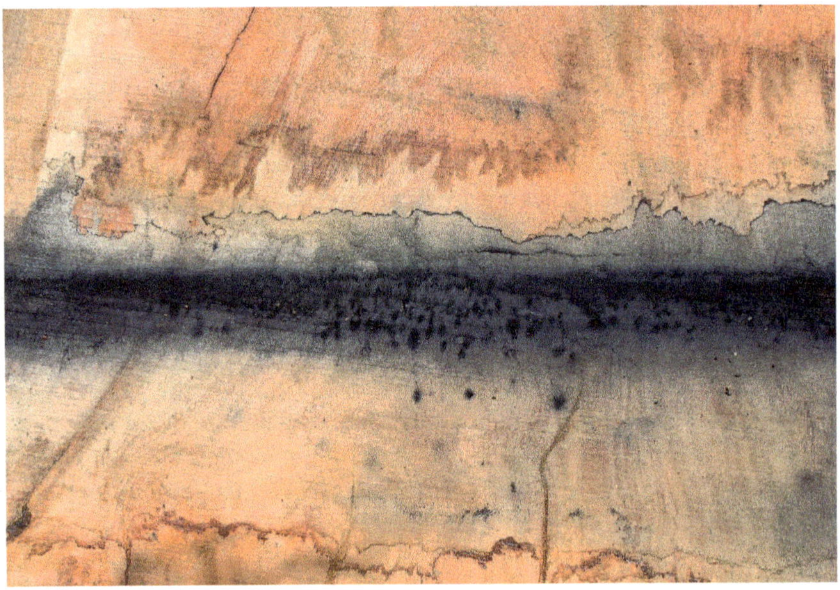

Surface of a freshly cut plank of *Eucalyptus camaldulensis* displaying thin-layer chromatography. The horizontal blue strip is from a reaction between the iron bandsaw supports and the acidic timber.

Separation of compounds is based on the competition of the solute and the mobile phase for binding places on the stationary phase. For instance, if normal-phase silica gel is used as the stationary phase, it can be considered polar. Given two compounds that differ in polarity, the more polar compound has a stronger interaction with the silica and is, therefore, more capable to dispel the mobile phase from the binding places. As a consequence, the less polar compound moves higher up the plate (resulting in a higher R_f value). If the mobile phase is changed to a more polar solvent or mixture of solvents, it is more capable of dispelling solutes from the silica binding places, and all compounds on the TLC plate will move higher up the plate. It is commonly said that "strong" solvents (eluents) push the analyzed compounds up the plate, whereas "weak" eluents barely move them. The order of strength/weakness depends on the coating (stationary phase) of the TLC plate. For silica gel-coated TLC plates, the eluent strength increases in the following order: perfluoroalkane (weakest), hexane, pentane, carbon tetrachloride, benzene/toluene, dichloromethane, diethyl ether, ethyl acetate, acetonitrile, acetone, 2-propanol/*n*-butanol, water, methanol, triethylamine, acetic acid, formic acid (strongest). For C18-coated plates the order is reverse. In other words, when the stationary phase is polar and the mobile phase is polar, the method is *normal-phase* as opposed to *reverse-phase*. This means that if a mixture of ethyl acetate and hexane as the mobile phase is used, adding more ethyl acetate results in higher R_f values for all compounds on the TLC plate. Changing the polarity of the mobile phase will normally not result in reversed order of running of the compounds on the TLC plate. An eluotropic series can be used as a guide in selecting a mobile phase. If a reversed order of running of the compounds is desired, an apolar stationary phase should be used, such as C18-functionalized silica.

Analysis

As the chemicals being separated may be colorless, several methods exist to visualize the spots:

- fluorescent analytes like quinine may be detected under blacklight (366 nm)
- Often a small amount of a fluorescent compound, usually manganese-activated zinc silicate, is added to the adsorbent that allows the visualization of spots under UV-C light (254 nm). The adsorbent layer will thus fluoresce light-green by itself, but spots of analyte quench this fluorescence.
- Iodine vapors are a general unspecific color reagent
- Specific color reagents into which the TLC plate is dipped or which are sprayed onto the plate exist.
- Potassium permanganate - oxidation
- Bromine
- In the case of lipids, the chromatogram may be transferred to a PVDF membrane and then subjected to further analysis, for example mass spectrometry, a technique known as Far-Eastern blotting.

Once visible, the R_f value, or retardation factor, of each spot can be determined by dividing the distance the product traveled by the distance the solvent front traveled using the initial spotting site as reference. These values depend on the solvent used and the type of TLC plate and are not physical constants.

Applications

Characterization

In organic chemistry, reactions are qualitatively monitored with TLC. Spots sampled with a capillary tube are placed on the plate: a spot of starting material, a spot from the reaction mixture, and a cross-spot with both. A small (3 by 7 cm) TLC plate takes a couple of minutes to run. The analysis is qualitative, and it will show if the starting material has disappeared, i.e. the reaction is complete, if any product has appeared, and how many products are generated (although this might be underestimated due to co-elution). Unfortunately, TLCs from low-temperature reactions may give misleading results, because the sample is warmed to room temperature in the capillary, which can alter the reaction—the warmed sample analyzed by TLC is not the same as what is in the low-temperature flask. One such reaction is the DIBALH reduction of ester to aldehyde.

In one study TLC has been applied in the screening of organic reactions, for example in the fine-tuning of BINAP synthesis from 2-naphthol. In this method, the alcohol and catalyst solution (for instance iron(III) chloride) are placed separately on the base line, then reacted, and then instantly analyzed. A special application of TLC is in the characterization of radiolabeled compounds, where it is used to determine radiochemical purity. The TLC sheet is visualized using a sheet of photographic film or an instrument capable of measuring radioactivity. It may be visualized using other means as well. This method is much more sensitive than the others and can be used to detect an extremely small amount of a compound, provided that it carries a radioactive atom.

Isolation

Since different compounds will travel a different distance in the stationary phase, chromatography can in effect be used as an isolation technique. The separated compounds each occupying a specific area on the plate, they can be scraped away, put in another solvent to separate them from the stationary phase and used for further analysis. As an example, in the chromatography of an extract of green leaves (for example spinach) in 7 stages of development, Carotene elutes quickly and is only visible until step 2. Chlorophyll A and B are halfway in the final step and lutein the first compound staining yellow. Once the chromatography is over, the carotene can be removed from the plate, put back into a solvent and ran into a spectrophotometer to characterize its wavelength absorption.

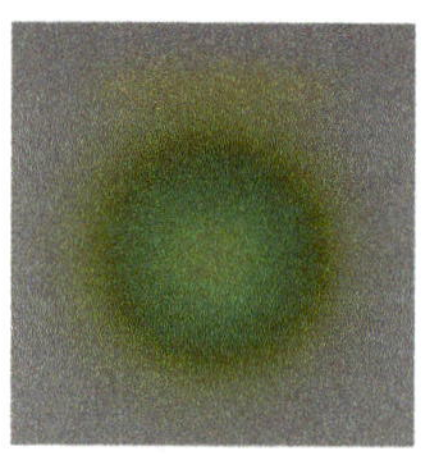

Step 1

Step 2

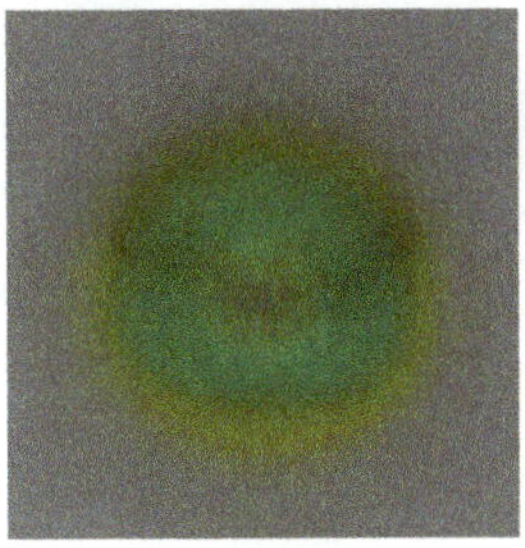

Step 3

Step 4

Step 5

Step 6

Step 7

HIGH-PERFORMANCE LIQUID CHROMATOGRAPHY

High-performance liquid chromatography (**HPLC**; formerly referred to as **high-pressure liquid chromatography**), is a technique in analytical chemistry used to separate, identify, and quantify each component in a mixture. It relies on pumps to pass a pressurized liquid solvent containing the sample mixture through a column filled with a solid adsorbent material. Each component in the sample interacts slightly differently with the adsorbent material, causing different flow rates for the different components and leading to the separation of the components as they flow out the column.

A modern self-contained HPLC.

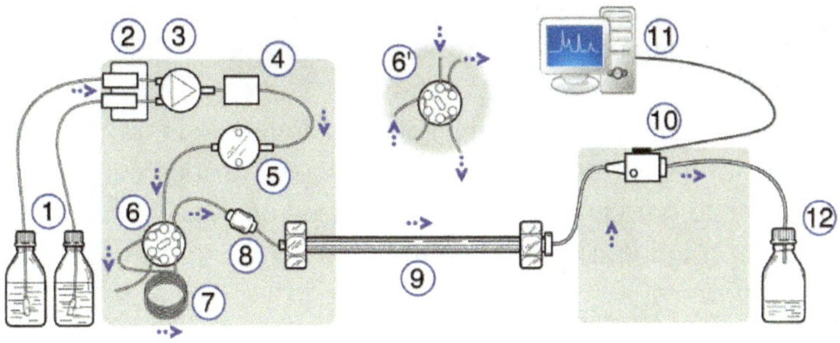

Schematic representation of an HPLC unit. (1) Solvent reservoirs, (2) Solvent degasser, (3) Gradient valve, (4) Mixing vessel for delivery of the mobile phase, (5) High-pressure pump, (6) Switching valve in "inject position", (6') Switching valve in "load position", (7) Sample injection loop, (8) Pre-column (guard column), (9) Analytical column, (10) Detector (i.e. IR, UV), (11) Data acquisition, (12) Waste or fraction collector.

HPLC has been used for manufacturing (e.g. during the production process of pharmaceutical and biological products), legal (e.g. detecting performance enhancement drugs in urine), research (e.g. separating the components of a complex biological sample, or of similar synthetic chemicals from each other), and medical (e.g. detecting vitamin D levels in blood serum) purposes.

Chromatography can be described as a mass transfer process involving adsorption. HPLC relies on pumps to pass a pressurized liquid and a sample mixture through a column filled with adsorbent, leading to the separation of the sample components. The active component of the column, the adsorbent, is typically a granular material made of solid particles (e.g. silica, polymers, etc.), 2–50 micrometers in size. The components of the sample mixture are separated from each other due to their different degrees of interaction with the adsorbent particles. The pressurized liquid is typically a mixture of solvents (e.g. water, acetonitrile and/or methanol) and is referred to as a mobile phase". Its composition and temperature play a major role in the separation process by influencing the interactions taking place between sample components and adsorbent. These interactions are physical in nature, such as hydrophobic (dispersive), dipole–dipole and ionic, most often a combination.

HPLC is distinguished from traditional (low pressure") liquid chromatography because operational pressures are significantly higher (50–350 bar), while ordinary liquid chromatography typically relies on the force of gravity to pass the mobile phase through the column. Due to the small sample amount separated in analytical HPLC, typical column dimensions are 2.1–4.6 mm diameter, and 30–250 mm length. Also HPLC columns are made with smaller sorbent particles (2–50 micrometer in average particle size). This gives HPLC superior resolving power (the ability to distinguish between compounds) when separating mixtures, which makes it a popular chromatographic technique.

The schematic of an HPLC instrument typically includes a sampler, pumps, and a detector. The sampler brings the sample mixture into the mobile phase stream which carries it into the column. The pumps deliver the desired flow and composition of the mobile phase through the column. The detector generates a signal proportional to the amount of sample component emerging from the column, hence allowing for quantitative analysis of the sample components. A digital microprocessor and user software control the HPLC instrument and provide data analysis. Some models of mechanical pumps in a HPLC instrument can mix multiple solvents together in ratios changing in time, generating a composition gradient in the mobile phase. Various detectors are in common use, such as UV/Vis, photodiode array (PDA) or based on mass spectrometry. Most HPLC instruments also have a column oven that allows for adjusting the temperature at which the separation is performed.

OPERATION

The sample mixture to be separated and analyzed is introduced, in a discrete small volume (typically microliters), into the stream of mobile phase percolating through the column. The components of the sample move through the column at different velocities, which are a function of specific physical interactions with the adsorbent (also called stationary phase). The velocity of each component depends on its chemical nature, on the nature of the stationary phase (column) and on the composition of the mobile phase. The time at which a specific analyte elutes (emerges from the column) is called its retention time. The retention time measured under particular conditions is an identifying characteristic of a given analyte.

Many different types of columns are available, filled with adsorbents varying in particle size, and in the nature of their surface ("surface chemistry"). The use of smaller particle size packing materials requires the use of higher operational pressure ("backpressure") and typically improves chromatographic resolution (i.e. the degree of separation between consecutive analytes emerging from the column). Sorbent particles may be hydrophobic or polar in nature.

Common mobile phases used include any miscible combination of water with various organic solvents (the most common are acetonitrile and methanol). Some HPLC techniques use water-free mobile phases (see Normal-phase chromatography below). The aqueous component of the mobile phase may contain acids (such as formic, phosphoric or trifluoroacetic acid) or salts to assist in the separation of the sample components. The composition of the mobile phase may be kept constant (isocratic elution mode") or varied (gradient elution mode") during the chromatographic analysis. Isocratic elution is typically effective in the separation of sample components that are not very different in their affinity for the stationary phase. In gradient elution the composition of the mobile phase is varied typically from low to high eluting strength. The eluting strength of the mobile phase is reflected by analyte retention times with high eluting strength producing fast elution (=short retention times). A typical gradient profile in reversed phase chromatography might start at 5% acctonitrile (in water or aqueous buffer) and progress linearly to 95% acetonitrile over 5–25 minutes. Periods of constant mobile phase composition may be part of any gradient profile. For example, the mobile phase composition may be kept constant at 5% acetonitrile for 1–3 min, followed by a linear change up to 95% acetonitrile.

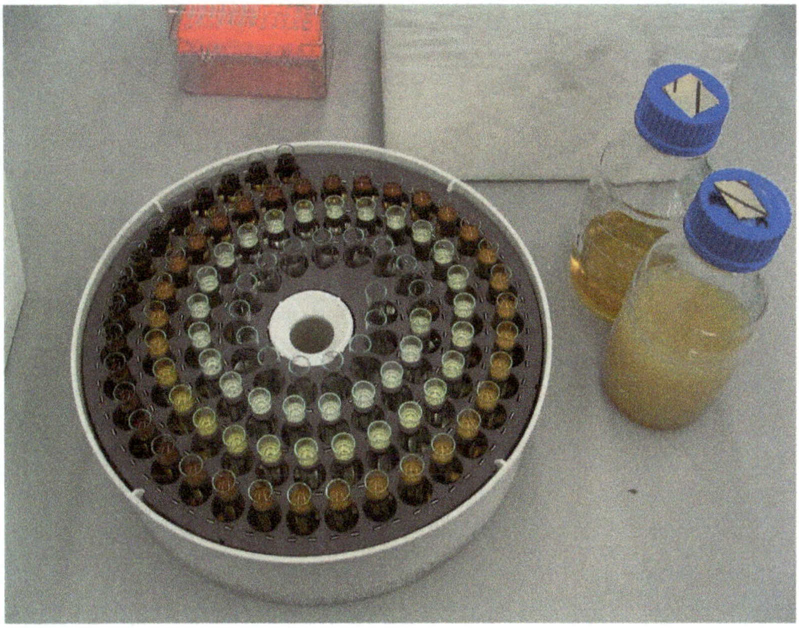

A rotary fraction collector collecting HPLC output. The system is being used to isolate a fraction containing Complex I from *E. coli* plasma membranes. About 50 litres of bacteria were needed to isolate this amount.

The chosen composition of the mobile phase (also called eluent) depends on the intensity of interactions between various sample components (analytes") and stationary phase (e.g. hydrophobic interactions in reversed-phase HPLC). Depending on their affinity for the stationary and mobile phases analytes partition between the two during the separation process taking place in the column. This partitioning process is similar to that which occurs during a liquid–liquid extraction but is continuous, not step-wise. In this example, using a water/acetonitrile gradient, more hydrophobic components will elute (come off the column) late, once the mobile phase gets more concentrated in acetonitrile (i.e. in a mobile phase of higher eluting strength).

The choice of mobile phase components, additives (such as salts or acids) and gradient conditions depends on the nature of the column and sample components. Often a series of trial runs is performed with the sample in order to find the HPLC method which gives adequate separation.

History and Development

Prior to HPLC scientists used standard liquid chromatographic techniques. Liquid chromatographic systems were largely inefficient due to the flow rate of solvents being dependent on gravity. Separations took many hours, and sometimes days to complete. Gas chromatography (GC) at the time was more powerful than liquid chromatography (LC), however, it was believed that gas phase separation and analysis of very polar high molecular weight biopolymers was impossible. GC was ineffective for many biochemists because of the thermal instability of the solutes.As a result, alternative methods were hypothesized which would soon result in the development of HPLC.

Following on the seminal work of Martin and Synge in 1941, it was predicted by Cal Giddings, Josef Huber, and others in the 1960s that LC could be operated in the high-efficiency mode by reducing the packing-particle diameter substantially below the typical LC (and GC) level of 150 μm and using pressure to increase the mobile phase velocity. These predictions underwent extensive experimentation and refinement throughout the 60s into the 70s. Early developmental research began to improve LC particles, and the invention of Zipax, a superficially porous particle, was promising for HPLC technology.

The 1970s brought about many developments in hardware and instrumentation. Researchers began using pumps and injectors to make a rudimentary design of an HPLC system. Gas amplifier pumps were ideal because they operated at constant pressure and did not require leak free seals or check valves for steady flow and good quantitation. Hardware milestones were made at Dupont IPD such as a low-dwell-volume gradient device being utilized as well as replacing the septum injector with a loop injection valve.

While instrumentational developments were important, the history of HPLC is primarily about the history and evolution of particle technology. After the introduction of porous layer particles, there has been a steady trend to reduced particle size to improve efficiency. However, by decreasing particle size, new problems arose. The practical disadvantages stem from the excessive pressure drop needed to force mobile fluid through the column and the difficulty of preparing a uniform packing of extremely fine materials. Every time particle size is reduced significantly, another round of instrument development usually must occur to handle the pressure.

Types

Partition Chromatography

Partition chromatography was one of the first kinds of chromatography that chemists developed. The partition coefficient principle has been applied in paper chromatography, thin layer chromatography, gas phase and liquid–liquid separation applications. The 1952 Nobel Prize in chemistry was earned by Archer John Porter Martin and Richard Laurence Millington Synge for their development of the technique, which was used for their separation of amino acids. Partition chromatography uses a retained solvent, on the surface or within the grains or fibers of an inert" solid supporting matrix as with paper chromatography; or takes advantage of some coulombic and/or hydrogen donor interaction with the stationary phase. Analyte molecules partition between a liquid stationary phase and the eluent. Just as in Hydrophilic Interaction Chromatography (HILIC; a sub-technique within HPLC), this method separates analytes based on differences in their polarity. HILIC most often uses a bonded polar stationary phase and a mobile phase made primarily of acetonitrile with water as the strong component. Partition HPLC has been used historically on unbonded silica or alumina supports. Each works effectively for separating analytes by relative polar differences. HILIC bonded phases have the advantage of separating acidic, basic and neutral solutes in a single chromatographic run.

HILIC is a Multi-Modal Partition Technique

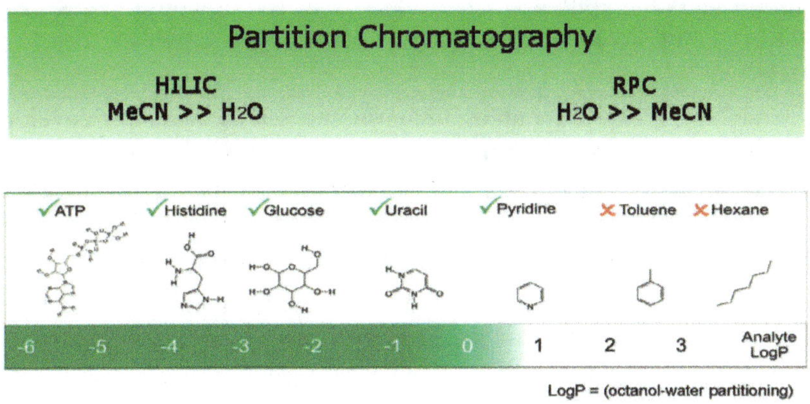

HILIC Partition Technique Useful Range.

The polar analytes diffuse into a stationary water layer associated with the polar stationary phase and are thus retained. The stronger the interactions between the polar analyte and the polar stationary phase (relative to the mobile phase) the longer the elution time. The interaction strength depends on the functional groups part of the analyte molecular structure, with more polarized groups (e.g. hydroxyl-) and groups capable of hydrogen bonding inducing more retention. Coulombic (electrostatic) interactions can also increase retention. Use of more polar solvents in the mobile phase will decrease the retention time of the analytes, whereas more hydrophobic solvents tend to increase retention times.

Normal–phase Chromatography

Normal–phase chromatography was one of the first kinds of HPLC that chemists developed. Also known as normal-phase HPLC (NP-HPLC) this method separates analytes based on their affinity for a polar stationary surface such as silica, hence it is based on analyte ability to engage in polar interactions (such as hydrogen-bonding or dipole-dipole type of interactions) with the sorbent surface. NP-HPLC uses a non-polar, non-aqueous mobile phase (e.g. Chloroform), and works effectively for separating analytes readily soluble in non-polar solvents. The analyte associates with and is retained by the polar stationary phase. Adsorption strengths increase with increased analyte polarity. The interaction strength depends not only on the functional groups present in the structure of the analyte molecule, but also on steric factors. The effect of steric hindrance on interaction strength allows this method to resolve (separate) structural isomers.

The use of more polar solvents in the mobile phase will decrease the retention time of analytes, whereas more hydrophobic solvents tend to induce slower elution (increased retention times). Very polar solvents such as traces of water in the mobile phase tend to adsorb to the solid surface of the stationary phase forming a stationary bound (water) layer which is considered to play an active role in retention. This behavior is somewhat peculiar to normal phase chromatography because it is governed almost exclusively by an adsorptive mechanism (i.e. analytes interact with a solid surface rather than with the solvated layer of a ligand attached to the sorbent surface; see also reversed-phase HPLC below). Adsorption chromatography is still widely used for structural isomer separations in both column and thin-layer chromatography formats on activated (dried) silica or alumina supports.

Partition- and NP-HPLC fell out of favor in the 1970s with the development of reversed-phase HPLC because of poor reproducibility of retention times due to the presence of a water or protic organic solvent layer on the surface of the silica or alumina chromatographic media. This layer changes with any changes in the composition of the mobile phase (e.g. moisture level) causing drifting retention times.

Recently, partition chromatography has become popular again with the development of Hilic bonded phases which demonstrate improved reproducibility, and due to a better understanding of the range of usefulness of the technique.

Displacement Chromatography

The basic principle of displacement chromatography is: A molecule with a high affinity for the chromatography matrix (the displacer) will compete effectively for binding sites, and thus displace all molecules with lesser affinities. There are distinct differences between displacement and elution chromatography. In elution mode, substances typically emerge from a column in narrow, Gaussian peaks. Wide separation of peaks, preferably to baseline, is desired in order to achieve maximum purification. The speed at which any component of a mixture travels down the column in elution mode depends on many factors. But for two substances to travel at different speeds, and thereby be resolved, there must be substantial differences in some interaction between the biomolecules and the chromatography matrix. Operating parameters are adjusted to maximize the effect of this difference. In many cases, baseline separation of the peaks can be achieved only with gradient elution and low column loadings. Thus, two drawbacks to elution mode chromatography, especially at the preparative scale, are operational complexity, due to gradient solvent pumping, and low throughput, due to low column loadings. Displacement chromatography has advantages over elution chromatography in that components are resolved into consecutive zones of pure substances rather than "peaks". Because the process takes advantage of the nonlinearity of the isotherms, a larger column feed can be separated on a given column with the purified components recovered at significantly higher concentration.

Reversed-Phase Chromatography (RPC)

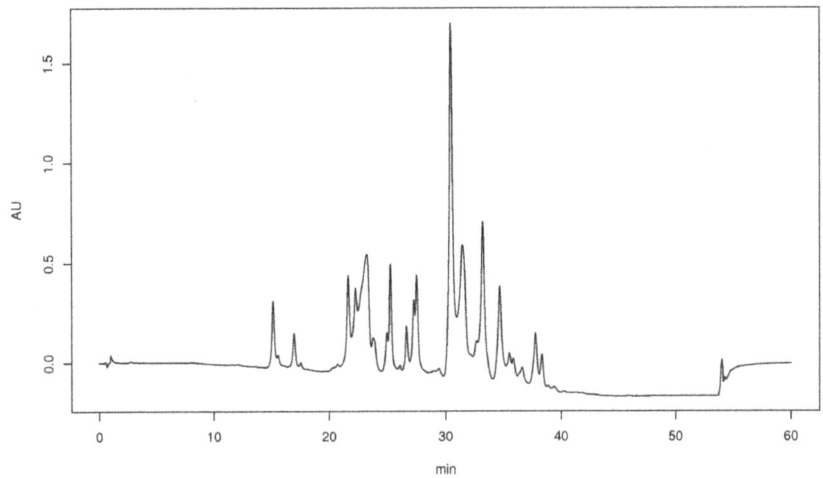

A chromatogram of complex mixture (perfume water) obtained by reversed phase HPLC.

Reversed phase HPLC (RP-HPLC) has a non-polar stationary phase and an aqueous, moderately polar mobile phase. One common stationary phase is a silica which has been surface-modified with RMe_2SiCl, where R is a straight chain alkyl group such as $C_{18}H_{37}$ or C_8H_{17}. With such stationary phases, retention time is longer for molecules which are less polar, while polar molecules elute more readily (early in the analysis). An investigator can increase retention times by adding more water to the mobile phase; thereby making the affinity of the hydrophobic analyte for the hydrophobic stationary phase stronger relative to the now more hydrophilic mobile phase. Similarly, an investigator can decrease retention time by adding more organic solvent to the eluent. RP-HPLC is so commonly used that it is often incorrectly referred to as "HPLC" without further specification. The pharmaceutical industry regularly employs RP-HPLC to qualify drugs before their release.

RP-HPLC operates on the principle of hydrophobic interactions, which originates from the high symmetry in the dipolar water structure and plays the most important role in all processes in life science. RP-HPLC allows the measurement of these interactive forces. The binding of the analyte to the stationary phase is proportional to the contact surface area around the non-polar segment of the analyte molecule upon association with the ligand on the stationary phase. This solvophobic effect is dominated by the force of water for "cavity-reduction" around the analyte and the C_{18}-chain versus the complex of both. The energy released in this process is proportional to the surface tension of the eluent (water: 7.3×10^{-6} J/cm^2, methanol: 2.2×10^{-6} J/cm^2) and to the hydrophobic surface of the analyte and the ligand respectively. The retention can be decreased by adding a less polar solvent (methanol, acetonitrile) into the mobile phase to reduce the surface tension of water. Gradient elution uses this effect by automatically reducing the polarity and the surface tension of the aqueous mobile phase during the course of the analysis.

Structural properties of the analyte molecule play an important role in its retention characteristics. In general, an analyte with a larger hydrophobic surface area (C–H, C–C, and generally non-polar atomic bonds, such as S-S and others) is retained longer because it is non-interacting with the water structure. On the other hand, analytes with higher polar surface area (conferred by the presence of polar groups, such as -OH, $-NH_2$, COO^- or $-NH_3^+$ in their structure) are less retained as they are better integrated into water. Such interactions are subject to steric effects in that very large molecules may have only restricted access to the pores of the stationary phase, where the interactions with surface ligands (alkyl chains) take place. Such surface hindrance typically results in less retention.

Retention time increases with hydrophobic (non-polar) surface area. Branched chain compounds elute more rapidly than their corresponding linear isomers because the overall surface area is decreased. Similarly organic compounds with single C–C bonds elute later than those with a C=C or C–C triple bond, as the double or triple bond is shorter than a single C–C bond.

Aside from mobile phase surface tension (organizational strength in eluent structure), other mobile phase modifiers can affect analyte retention. For example, the addition of inorganic salts causes a moderate linear increase in the surface tension of aqueous solutions (ca. 1.5×10^{-7} J/cm² per Mol for NaCl, 2.5×10^{-7} J/cm² per Mol for $(NH_4)_2SO_4$), and because the entropy of the analyte-solvent interface is controlled by surface tension, the addition of salts tend to increase the retention time. This technique is used for mild separation and recovery of proteins and protection of their biological activity in protein analysis (hydrophobic interaction chromatography, HIC).

Another important factor is the mobile phase pH since it can change the hydrophobic character of the analyte. For this reason most methods use a buffering agent, such as sodium phosphate, to control the pH. Buffers serve multiple purposes: control of pH, neutralize the charge on the silica surface of the stationary phase and act as ion pairing agents to neutralize analyte charge. Ammonium formate is commonly added in mass spectrometry to improve detection of certain analytes by the formation of analyte-ammonium adducts. A volatile organic acid such as acetic acid, or most commonly formic acid, is often added to the mobile phase if mass spectrometry is used to analyze the column effluent. Trifluoroacetic acid is used infrequently in mass spectrometry applications due to its persistence in the detector and solvent delivery system, but can be effective in improving retention of analytes such as carboxylic acids in applications utilizing other detectors, as it is a fairly strong organic acid. The effects of acids and buffers vary by application but generally improve chromatographic resolution.

Reversed phase columns are quite difficult to damage compared with normal silica columns; however, many reversed phase columns consist of alkyl derivatized silica particles and should **never** be used with aqueous bases as these will destroy the underlying silica particle. They can be used with aqueous acid, but the column should not be exposed to the acid for too long, as it can corrode the metal parts of the HPLC equipment. RP-HPLC columns should be flushed with clean solvent after use to remove residual acids or buffers, and stored in an appropriate composition of solvent. The metal content of HPLC columns must be kept low if the best possible ability to separate substances is to be retained. A good test for the metal content of a column is to inject a sample which is a mixture of 2,2`- and 4,4`- bipyridine. Because the 2,2`-bipy can chelate the metal, the shape of the peak for the 2,2'-bipy will be distorted (tailed) when metal ions are present on the surface of the silica...

Size-Exclusion Chromatography

Size-exclusion chromatography (SEC), also known as *gel permeation chromatography* or *gel filtration chromatography*, separates particles on the basis of molecular size (actually by a particle's Stokes radius). It is generally a low resolution chromatography and thus it is often reserved for the final, "polishing" step of the purification. It is also useful for determining the tertiary structure and quaternary structure of purified proteins. SEC is used primarily for the analysis of large molecules such as proteins or polymers. SEC works by trapping these smaller molecules in the pores of a particle. The larger molecules simply pass by the pores as they are too large to enter the pores. Larger molecules therefore flow through the column quicker than smaller molecules, that is, the smaller the molecule, the longer the retention time.

This technique is widely used for the molecular weight determination of polysaccharides. SEC is the official technique (suggested by European pharmacopeia) for the molecular weight comparison of different commercially available low-molecular weight heparins.

Ion-exchange Chromatography

In ion-exchange chromatography (IC), retention is based on the attraction between solute ions and charged sites bound to the stationary phase. Solute ions of the same charge as the charged sites on the column are excluded from binding, while solute ions of the opposite charge of the charged sites of the column are retained on the column. Solute ions that are retained on the column can be eluted from the column by changing the solvent conditions (e.g. increasing the ion effect of the solvent system by increasing the salt concentration of the solution, increasing the column temperature, changing the pH of the solvent, etc...).

Types of ion exchangers include:

- Polystyrene resins – These allow cross linkage which increases the stability of the chain. Higher cross linkage reduces swerving, which increases the equilibration time and ultimately improves selectivity.
- Cellulose and dextran ion exchangers (gels) – These possess larger pore sizes and low charge densities making them suitable for protein separation.
- Controlled-pore glass or porous silica

In general, ion exchangers favor the binding of ions of higher charge and smaller radius.

An increase in counter ion (with respect to the functional groups in resins) concentration reduces the retention time. A decrease in pH reduces the retention time in cation exchange while an increase in pH reduces the retention time in anion exchange. By lowering the pH of the solvent in a cation exchange column, for instance, more hydrogen ions are available to compete for positions on the anionic stationary phase, thereby eluting weakly bound cations.

This form of chromatography is widely used in the following applications: water purification, preconcentration of trace components, ligand-exchange chromatography, ion-exchange chromatography of proteins, high-pH anion-exchange chromatography of carbohydrates and oligosaccharides, and others.

Bioaffinity Chromatography

This chromatographic process relies on the property of biologically active substances to form stable, specific, and reversible complexes. The formation of these complexes involves the participation of common molecular forces such as the Van der Waals interaction, electrostatic interaction, dipole-dipole interaction, hydrophobic interaction, and the hydrogen bond. An efficient, biospecific bond is formed by a simultaneous and concerted action of several of these forces in the complementary binding sites.

Aqueous normal-phase chromatography

Aqueous normal-phase chromatography (ANP) is a chromatographic technique which encompasses the mobile phase region between reversed-phase chromatography (RP) and organic normal phase chromatography (ONP). This technique is used to achieve unique selectivity for hydrophilic compounds, showing normal phase elution using reversed-phase solvents.

ISOCRATIC AND GRADIENT ELUTION

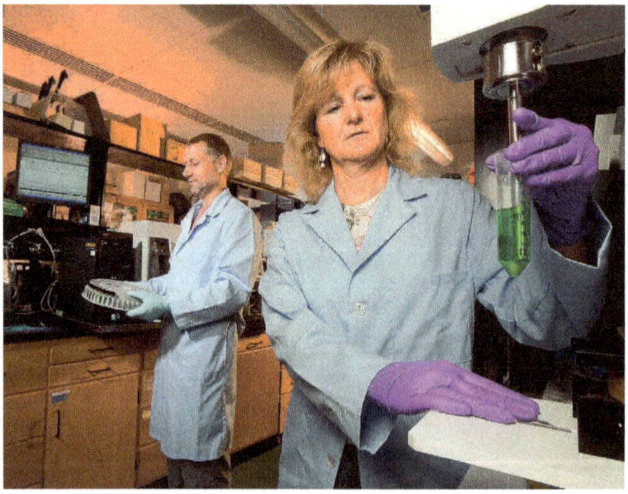

At the ARS Natural Products Utilization Research Unit in Oxford, MS., a support scientist (r) extracts plant pigments that will be analyzed by a plant physiologist (l) using an HPLC system.

A separation in which the mobile phase composition remains constant throughout the procedure is termed *isocratic* (meaning *constant composition*). The word was coined by Csaba Horvath who was one of the pioneers of HPLC.,

The mobile phase composition does not have to remain constant. A separation in which the mobile phase composition is changed during the separation process is described as a *gradient elution*. One example is a gradient starting at 10% methanol and ending at 90% methanol after 20 minutes. The two components of the mobile phase are typically termed "A" and "B"; *A* is the "weak" solvent which allows the solute to elute only slowly, while *B* is the "strong" solvent which rapidly elutes the solutes from the column. In reversed-phase chromatography, solvent *A* is often water or an aqueous buffer, while *B* is an organic solvent miscible with water, such as acetonitrile, methanol, THF, or isopropanol.

In isocratic elution, peak width increases with retention time linearly according to the equation for N, the number of theoretical plates. This leads to the disadvantage that late-eluting peaks get very flat and broad. Their shape and width may keep them from being recognized as peaks.

Gradient elution decreases the retention of the later-eluting components so that they elute faster, giving narrower (and taller) peaks for most components. This also improves the peak shape for tailed peaks, as the increasing concentration of the organic eluent pushes the tailing part of a peak forward. This also increases the peak height (the peak looks sharper"), which is important in trace analysis. The gradient program may include sudden step" increases in the percentage of the organic component, or different slopes at different times – all according to the desire for optimum separation in minimum time.

In isocratic elution, the selectivity does not change if the column dimensions (length and inner diameter) change – that is, the peaks elute in the same order. In gradient elution, the elution order may change as the dimensions or flow rate change.

The driving force in reversed phase chromatography originates in the high order of the water structure. The role of the *organic component of the mobile phase* is to reduce this high order and thus *reduce the retarding strength of the aqueous component.*

PARAMETERS

Theoretical

HPLC separations have theoretical parameters and equations to describe the separation of components into signal peaks when detected by instrumentation such as by a UV detector or a mass spectrometer. The parameters are largely derived from two sets of chromatagraphic theory: plate theory (as part of Partition chromatography), and the rate theory of chromatography / *Van Deemter equation*. Of course, they can be put in practice through analysis of HPLC chromatograms, although rate theory is considered the more accurate theory.

They are analogous to the calculation of retention factor for a paper chromatography separation, but describes how well HPLC separates a mixture into two or more components that are detected as peaks (bands) on a chromatogram. The HPLC parameters are the:

efficiency factor(*N*), the retention factor (kappa prime), and the separation factor (alpha). Together the factors are variables in a resolution equation, which describes how well two components' peaks separated or overlapped each other. These parameters are mostly only used for describing HPLC reversed phase and HPLC normal phase separations, since those separations tend to be more subtle than other HPLC modes (e.g. ion exchange and size exclusion).

- Void volume is the amount of space in a column that is occupied by solvent. It is the space within the column that is outside of the column's internal packing material. Void volume is measured on a chromatogram as the first component peak detected, which is usually the solvent that was present in the sample mixture; ideally the sample solvent flows through the column without interacting with the column, but is still detectable as distinct from the HPLC solvent. The void volume is used as a correction factor.
- Efficiency factor (*N*) practically measures how sharp component peaks on the chromatogram are, as ratio of the component peak's area ("retention time") relative to the width of the peaks at their widest point (at the baseline). Peaks that are tall, sharp, and relatively narrow indicate that separation method efficiently removed a component from a mixture; high efficiency. Efficiency is very dependent upon the HPLC column and the HPLC method used. Efficiency factor is synonymous with plate number, and the 'number of theoretical plates'.
- Retention factor (*kappa prime*) measures how long a component of the mixture stuck to the column, measured by the area under the curve of its peak in a chromatogram (since HPLC chromatograms are a function of time). Each chromatogram peak will have its own retention factor (e.g. *kappa*1 for the retention factor of the first peak). This factor may be corrected for by the void volume of the column.
- Separation factor (*alpha*) is a relative comparison on how well two neighboring components of the mixture were separated (i.e. two neighboring bands on a chromatogram). This factor is defined in terms of a ratio of the retention factors of a pair of neighboring chromatogram peaks, and may also be corrected for by the void volume of the column. The greater the separation factor value is over 1.0, the better the separation, until about 2.0 beyond which an HPLC method is probably not needed for separation.

Resolution equations relate the three factors such that high efficiency and separation factors improve the resolution of component peaks in a HPLC separation.

Internal Diameter

The internal diameter (ID) of an HPLC column is an important parameter that influences the detection sensitivity and separation selectivity in gradient elution. It also determines the quantity of analyte that can be loaded onto the column. Larger columns are usually seen in industrial applications, such as the purification of a drug product for later use. Low-ID columns have improved sensitivity and lower solvent consumption at the expense of loading capacity.

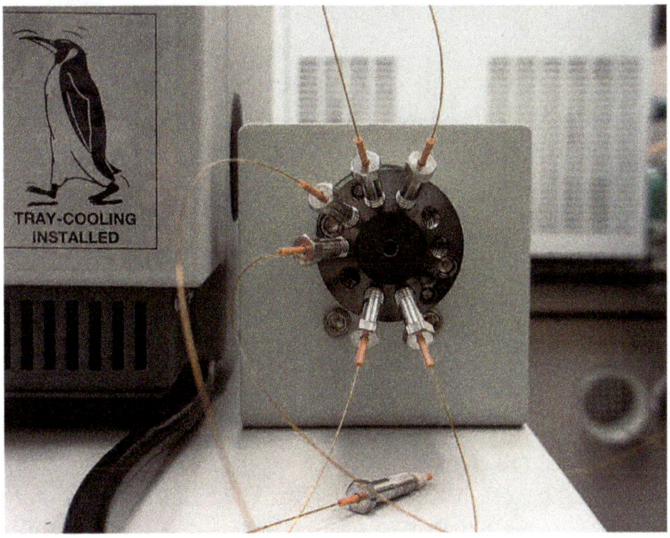

Tubing on a nano-liquid chromatography (nano-LC) system, used for very low flow capacities.

- Larger ID columns (over 10 mm) are used to purify usable amounts of material because of their large loading capacity.
- Analytical scale columns (4.6 mm) have been the most common type of columns, though smaller columns are rapidly gaining in popularity. They are used in traditional quantitative analysis of samples and often use a UV-Vis absorbance detector.
- Narrow-bore columns (1–2 mm) are used for applications when more sensitivity is desired either with special UV-vis detectors, fluorescence detection or with other detection methods like liquid chromatography-mass spectrometry
- Capillary columns (under 0.3 mm) are used almost exclusively with alternative detection means such as mass spectrometry. They are usually made from fused silica capillaries, rather than the stainless steel tubing that larger columns employ.

Particle Size

Most traditional HPLC is performed with the stationary phase attached to the outside of small spherical silica particles (very small beads). These particles come in a variety of sizes with 5 μm beads being the most common. Smaller particles generally provide more surface area and better separations, but the pressure required for optimum linear velocity increases by the inverse of the particle diameter squared.

This means that changing to particles that are half as big, keeping the size of the column the same, will double the performance, but increase the required pressure by a factor of four. Larger particles are used in preparative HPLC (column diameters 5 cm up to >30 cm) and for non-HPLC applications such as solid-phase extraction.

Pore Size

Many stationary phases are porous to provide greater surface area. Small pores provide greater surface area while larger pore size has better kinetics, especially for larger analytes. For example, a protein which is only slightly smaller than a pore might enter the pore but does not easily leave once inside.

Pump Pressure

Pumps vary in pressure capacity, but their performance is measured on their ability to yield a consistent and reproducible flow rate. Pressure may reach as high as 60 MPa (6000 lbf/in^2), or about 600 atmospheres. Modern HPLC systems have been improved to work at much higher pressures, and therefore are able to use much smaller particle sizes in the columns (<2 μm). These Ultra High Performance Liquid Chromatography" systems or UHPLCs can work at up to 120 MPa (17,405 lbf/in^2), or about 1200 atmospheres. The term "UPLC" is a trademark of the Waters Corporation, but is sometimes used to refer to the more general technique of UHPLC.

Detectors

HPLC most commonly uses a UV-Vis absorbance detector, however, a wide range of other chromatography detectors can be used. A kind of commonly utilized detector includes refractive index detectors, which provide readings by measuring the changes in the refractive index of the effluent as it moves through the flow cell. In certain cases, it is possible to use multiple detectors, for example LCMS normally combines UV-Vis with a mass spectrometer.

Autosamplers

Large numbers of samples can be automatically injected onto an HPLC system, by the use of HPLC autosamplers. In addition, HPLC autosamplers have an injection volume and technique which is exactly the same for each injection, consequently they provide a high degree of injection volume precision.

APPLICATIONS

Manufacturing

As briefly mentioned, HPLC has many applications in both laboratory and clinical science. It is a common technique used in pharmaceutical development as it is a dependable way to obtain and ensure product purity. While HPLC can produce extremely high quality (pure) products, it is not always the primary method used in the production of bulk drug materials. According to the European pharmacopoeia, HPLC is used in only 15.5% of syntheses. However, it plays a role in 44% of syntheses in the United States pharmacopoeia.

This could possibly be due to differences in monetary and time constraints, as HPLC on a large scale can be an expensive technique. An increase in specificity, precision, and accuracy that occurs with HPLC unfortunately corresponds to an increase in cost.

Legal

This technique is also used for detection of illicit drugs in urine. The most common method of drug detection is an immunoassay. This method is much more convenient. However, convenience comes at the cost of specificity and coverage of a wide range of drugs. As HPLC is a method of determining (and possibly increasing) purity, using HPLC alone in evaluating concentrations of drugs is somewhat insufficient. With this, HPLC in this context is often performed in conjunction with mass spectrometry. Using liquid chromatography instead of gas chromatography in conjunction with MS circumvents the necessity for derivitizing with acetylating or alkylation agents, which can be a burdensome extra step. This technique has been used to detect a variety of agents like doping agents, drug metabolites, glucuronide conjugates, amphetamines, opioids, cocaine, BZDs, ketamine, LSD, cannabis, and pesticides Performing HPLC in conjunction with Mass spectrometry reduces the absolute need for standardizing HPLC experimental runs.

Research

Similar assays can be performed for research purposes, detecting concentrations of potential clinical candidates like anti-fungal and asthma drugs. This technique is obviously useful in observing multiple species in collected samples, as well, but requires the use of standard solutions when information about species identity is sought out. It is used as a method to confirm results of synthesis reactions, as purity is essential in this type of research. However, mass spectrometry is still the more reliable way to identify species.

Medical

Medical use of HPLC can include drug analysis, but falls more closely under the category of nutrient analysis. While urine is the most common medium for analyzing drug concentrations, blood serum is the sample collected for most medical analyses with HPLC. Other methods of detection of molecules that are useful for clinical studies have been tested against HPLC, namely immunoassays. In one example of this, competitive protein binding assays (CPBA) and HPLC were compared for sensitivity in detection of vitamin D. Useful for diagnosing vitamin D deficiencies in children, it was found that sensitivity and specificity of this CPBA reached only 40% and 60%, respectively, of the capacity of HPLC. While an expensive tool, the accuracy of HPLC is nearly unparalleled.

FAST PROTEIN LIQUID CHROMATOGRAPHY

Fast protein liquid chromatography (FPLC), is a form of liquid chromatography that is often used to analyze or purify mixtures of proteins. As in other forms of chromatography, separation is possible because the different components of a mixture have different affinities for two materials, a moving fluid (the "mobile phase") and a porous solid (the stationary phase). In FPLC the mobile phase is an aqueous solution, or "buffer". The buffer flow rate is controlled by a positive-displacement pump and is normally kept constant, while the composition of the buffer can be varied by drawing fluids in different proportions from two or more external reservoirs. The stationary phase is a resin composed of beads, usually of cross-linked agarose, packed into a cylindrical glass or plastic column. FPLC resins are available in a wide range of bead sizes and surface ligands depending on the application.

In the most common FPLC strategy, ion exchange, a resin is chosen that the protein of interest will bind to the resin by a charge interaction while in buffer A (the running buffer) but become dissociated and return to solution in buffer B (the elution buffer). A mixture containing one or more proteins of interest is dissolved in 100% buffer A and pumped into the column. The proteins of interest bind to the resin while other components are carried out in the buffer. The total flow rate of the buffer is kept constant; however, the proportion of Buffer B (the elution" buffer) is gradually increased from 0% to 100% according to a programmed change in concentration (the gradient"). At some point during this process each of the bound proteins dissociates and appears in the effluent. The effluent passes through two detectors which measure salt concentration (by conductivity) and protein concentration (by absorption of ultraviolet light at a wavelength of 280nm). As each protein is eluted it appears in the effluent as a peak" in protein concentration and can be collected for further use.

FPLC was developed and marketed in Sweden by Pharmacia in 1982 and was originally called **fast performance liquid chromatography** to contrast it with HPLC or high-performance liquid chromatography. FPLC is generally applied only to proteins; however, because of the wide choice of resins and buffers it has broad applications. In contrast to HPLC the buffer pressure used is relatively low, typically less than 5 bar, but the flow rate is relatively high, typically 1-5 ml/min. FPLC can be readily scaled from analysis of milligrams of mixtures in columns with a total volume of 5ml or less to industrial production of kilograms of purified protein in columns with volumes of many liters. When used for analysis of mixtures the effluent is usually collected in fractions of 1-5 ml which can be further analyzed, e.g. by MALDI mass spectrometry. When used for protein purification there may be only two collection containers, one for the purified product and one for waste.

FPLC System Components

A typical laboratory FPLC consist of one or two high-precision pumps, a control unit, a column, a detection system and a fraction collector. Although it is possible to operate the system manually, the components are normally linked to a personal computer or, in older units, a microcontroller.

- Pumps: Majority of systems utilize two two-cylinder piston pumps, one for each buffer, combining the output of both in a mixing chamber. Some simpler systems use a single peristaltic pump which draws both buffers from separate reservoirs through a proportioning valve and mixing chamber. In either case the system allows the fraction of each buffer entering the column to be continuously varied. The flow rate can go from a few milliliters per minute in bench-top systems to liters per minute for industrial scale purifications. The wide flow range makes it suitable both for analytical and preparative chromatography.

- Injection loop: A segment of tubing of known volume which is filled with the sample solution before it is injected into the column. Loop volume can range from a few microliters to 50ml or more.
- Injection valve: A motorized valve which links the mixer and sample loop to the column. Typically the valve has three positions for loading the sample loop, for injecting the sample from the loop into the column, and for connecting the pumps directly to the waste line to wash them or change buffer solutions. The injection valve has a sample loading port through which the sample can be loaded into the injection loop, usually from a hypodermic syringe using a Luer-lock connection.
- Column: The column is a glass or plastic cylinder packed with beads of resin and filled with buffer solution. It is normally mounted vertically with the buffer flowing downward from top to bottom. A glass frit at the bottom of the column retains the resin beads in the column while allowing the buffer and dissolved proteins to exit.
- Flow Cells: The effluent from the column passes through one or more flow cells to measure the concentration of protein in the effluent (by UV light absorption at 280nm). The conductivity cell measures the buffer conductivity, usually in millisiemens/cm, which indicates the concentration of salt in the buffer. A flow cell which measures pH of the buffer is also commonly included. Usually each flow cell is connected to a separate electronics module which provides power and amplifies the signal.
- Monitor/Recorder: The flow cells are connected to a display and/or recorder. On older systems this was a simple chart recorder, on modern systems a computer with hardware interface and display is used. This permits the experimenter to identify when peaks in protein concentration occur, indicating that specific components of the mixture are being eluted.
- Fraction collector: The fraction collector is typically a rotating rack that can be filled with test tubes or similar containers. It allows samples to be collected in fixed volumes, or can be controlled to direct specific fractions detected as peaks of protein concentration, into separate containers.

Many systems include various optional components. A filter may be added between the mixer and column to minimize clogging. In large FPLC columns the sample may be loaded into the column directly using a small peristaltic pump rather than an injection loop. When the buffer contains dissolved gas, bubbles may form as pressure drops where the buffer exits the column; these bubbles create artifacts if they pass through the flow cells. This may be prevented by degassing the buffers, e.g. with a degasser, or by adding a flow restrictor downstream of the flow cells to maintain a pressure of 1-5 bar in the effluent line.

FPLC Columns

The columns used in FPLC are large [mm id] tubes that contain small [μ] particles or gel beads that are known as stationary phase. The chromatographic bed is composed by the gel beads inside the column and the sample is introduced into the injector and carried into the column by the flowing solvent. As a result of different components adhering to or diffusing through the gel, the sample mixture gets separated.

Columns used with an FPLC can separate macromolecules based on size, charge distribution (ion exchange), hydrophobicity, reverse-phase or biorecognition (as with affinity chromatography). For easy use, a wide range of pre-packed columns for techniques such as ion exchange, gel filtration (size exclusion), hydrophobic interaction, and affinity chromatography are available. FPLC differs from HPLC in that the columns used for FPLC can only be used up to maximum pressure of 3-4 MPa (435-580 psi). Thus, if the pressure of HPLC can be limited, each FPLC column may also be used in an HPLC machine.

Optimizing Protein Purification

Using a combination of chromatographic methods, purification of the target molecule is achieved. The purpose of purifying proteins with FPLC is to deliver quantities of the target at sufficient purity in a biologically active state to suit its further use. The quality of the end product varies depending the type and amount of starting material, efficiency of separation, and selectivity of the purification resin. The ultimate goal of a given purification protocol is to deliver the required yield and purity of the target molecule in the quickest, cheapest, and safest way for acceptable results. The range of purity required can be from that required for basic analysis (SDS-PAGE or ELISA, for example), with only bulk impurities removed, to pure enough for structural analysis (NMR or X-ray crystallography), approaching >99% target molecule. Purity required can also mean pure enough that the biological activity of the target is retained. These demands can be used to determine the amount of starting material required to reach the experimental goal. If the starting material is limited and full optimization of purification protocol cannot be performed, then a safe standard protocol that requires a minimum adjustment and optimization steps are expected. This may not be optimal with respect to experimental time, yield, and economy but it will achieve the experimental goal. On the other hand, if the starting material is enough to develop more complete protocol, the amount of work to reach the separation goal depends on the available sample information and target molecule properties. Limits to development of purification protocols many times depends on the source of the substance to be purified, whether from natural sources (harvested tissues or organisms, for example), recombinant sources (such as using prokaryotic or eukaryotic vectors in their respective expression systems), or totally synthetic sources.

No chromatographic techniques provide 100% yield of active material and overall yields depend on the number of steps in the purification protocol. By optimizing each step for the intended purpose and arranging them that minimizes inter step treatments, the number of steps will be minimized.

A typical multistep purification protocol starts with a preliminary capture step which many times utilizes ion exchange chromatography (IEC). The media (stationary phase) employed range from large bead resins (good for fast flow rates and little to no sample clarification at the expense of resolution) to small bead resins (for best possible resolution with all other factors being equal). Short and wide column geometries are amenable to high flow rates also at the expense of resolution, typically because of lateral diffusion of sample on the column. For techniques such as size exclusion chromatography to be useful, very long, thin columns and minimal sample volumes (maximum 5% of column volume) are required. Hydrophobic interaction chromatography (HIC) can also be used for first and/ or intermediate steps. Selectivity in HIC is independent of running pH and descending salt gradients are used. For HIC, conditioning involves adding ammonium sulphate to the sample to match the buffer A concentration. If HIC is used before IEC, the ionic strength would have to be lowered to match that of buffer A for IEC step by dilution, dialysis or buffer exchange by gel filtration. This is why IEC is usually performed prior to HIC as the high salt elution conditions for IEC are ideal for binding to HIC resins in the next purification step. Polishing is used to achieve the final level of purification required and is commonly performed on a gel filtration column. An extra intermediate purification step can be added or optimization of the different steps is performed for improving purity. This extra step usually involves another round of IEC under completely different conditions.

Although this is an example of a common purification protocol for proteins, the buffer conditions, flow rates, and resins used to achieve final goals can be chosen to cover a broad range of target proteins. This flexibility is imperative for a functional purification system as all proteins behave differently and often deviate from predictions.

REFERENCES

1. Gerber, F.; Krummen, M.; Potgeter, H.; Roth, A.; Siffrin, C.; Spoendlin, C. (2004). Practical aspects of fast reversed-phase high-performance liquid chromatography using 3μm particle packed columns and monolithic columns in pharmaceutical development and production working under current good manufacturing practice". Journal of Chromatography A. **1036** *(2): 127–133. doi:10.1016/j.chroma.2004.02.056. PMID 15146913.*
2. Morgan, David J. Fraction collector (post on Flickr)". Flickr. Retrieved 28 October 2015.
3. Karger, Barry L. (1997). HPLC: Early and Recent Perspectives". Journal of Chemical Education. **74**: 45. doi:10.1021/ed074p45.
4. Henry, Richard A. (1 February 2009) "The Early Days of HPLC at Dupont". Chromatography Online. Avanstar Communications Inc.
5. Iler, R.K. (1979) The Chemistry of Silica. John Wiley & Sons. New York.
6. Karger, B. L.; Berry, L. V. (1971). Rapid liquid-chromatographic separation of steroids on columns heavily loaded with stationary phase". Clin. Chem. **17** *(8): 757–64. PMID 4254537.*
7. Giddings, J. Calvin (1965) Dynamics of Chromatography, Part I. Principles

and Theory. Marcel Dekker, Inc., New York. p. 281.

8. Ettre, C. (2001). Milestones in Chromatography: The Birth of Partition Chromatography" (PDF). LCGC. **19** *(5): 506–512. Retrieved 2016-02-26.*

9. Martin, A J P; Synge, R L M (1941). Separation of the higher monoamino-acids by counter-current liquid-liquid extraction: the amino-acid composition of wool". Biochemical Journal. **35** *(1–2): 91–121.* doi:10.1042/bj0350091. PMC 1265473. PMID 16747393.

10. Lindsay, S.; Kealey, D. (1987). High performance liquid chromatography. Wiley. from review Hung, L. B.; Parcher, J. F.; Shores, J. C.; Ward, E. H. (1988). Theoretical and experimental foundation for surface-coverage programming in gas–solid chromatography with an adsorbable carrier gas". J. Am. Chem. Soc. **110** *(11): 1090–1096.* doi:10.1021/ac00162a003.

11. Displacement Chromatography. Sacheminc.com. Retrieved 2011-06-07. Archived September 15, 2008, at the Wayback Machine.

12. Snyder, Lloyd R.; Dolan, John W. (2006). High-Performance Gradient Elution: The Practical Application of the Linear-Solvent-Strength Model. Wiley Interscience. ISBN 0470055510.

13. Majors, Ronald E.. (2010-09-07) Fast and Ultrafast HPLC on sub-2 μm Porous Particles — Where Do We Go From Here? – LC-GC Europe. Lcgceurope.com. Retrieved 2011-06-07.

14. Xiang, Y.; Liu Y.; Lee M.L. (2006). Ultrahigh pressure liquid chromatography using elevated temperature". Journal of Chromatography A. **1104** *(1–2): 198–202. doi:10.1016/j.chroma.2005.11.118. PMID 16376355.*

15. Horváth, Cs.; Preiss B.A.; Lipsky S.R. (1967). Fast liquid chromatography. Investigation of operating parameters and the separation of nucleotides on pellicular ion exchangers". Analytical Chemistry. **39** *(12): 1422–1428.* doi:10.1021/ac60256a003. PMID 6073805.

16. 1290 Infinity Quaternary Pump. Agilent

17. Gerber, Frederic (May 2004). Practical aspects of fast reversed-phase high-performance liquid chromatography using 3 *μm* particle packed columns and monolithic columns in pharmaceutical development and production working under current good manufacturing practice". Journal of Chromatography. **1036** *(2): 127–33. doi:10.1016/j.chroma.2004.02.056. PMID 15146913.*

18. Siddiqui, Masoom Raza; AlOthman, Zeid A.; Rahman, Nafisur (2013). Analytical techniques in pharmaceutical analysis: A review". Arabian Journal of Chemistry. doi:10.1016/j.arabjc.2013.04.016.

19. The European Pharmacopoeia, 2002. fourth ed., Council of Europe, Strasbourg.

20. United States Pharmacopoeia, 2004. 27th ed. The USP Convention Inc., Rockville, MD.

21. Pesce, Amadeo; Rosenthal, Murray; West, Robert; West, Cameron; Crews, Bridgit; Mikel, Charles; Almazan, Perla; Latyshev, Sergey (2010-06-01). An evaluation of the diagnostic accuracy of liquid chromatography-tandem mass

spectrometry versus immunoassay drug testing in pain patients". Pain Physician. **13** *(3): 273–281. PMID 20495592.*

22. Tsai, I.-Lin; Weng, Te-I.; Tseng, Yufeng J.; Tan, Happy Kuy-Lok; Sun, Hsiao-Ju; Kuo, Ching-Hua (2013-12-01). Screening and confirmation of 62 drugs of abuse and metabolites in urine by ultra-high-performance liquid chromatography-quadrupole time-of-flight mass spectrometry". Journal of Analytical Toxicology. **37** *(9): 642–651. doi:10.1093/*jat/bkt083. PMID 24084874.
23. Weinmann, W.; Renz, M.; Vogt, S.; Pollak, S. (2000-01-01). Automated solid-phase extraction and two-step derivatisation for simultaneous analysis of basic illicit drugs in serum by GC/MS". International Journal of Legal Medicine. **113** *(4): 229–235.* doi:10.1007/s004149900098. PMID 10929239.
24. Kolmonen, Marjo; Leinonen, Antti; Pelander, Anna; Ojanperä, Ilkka (2007-02-28). A general screening method for doping agents in human urine by solid phase extraction and liquid chromatography/time-of-flight mass spectrometry". Analytica Chimica Acta. **585** *(1): 94–102. doi:10.1016/j.aca.2006.12.028. PMID 17386652.*
25. Pelander, Anna; Ojanperä, Ilkka; Laks, Suvi; Rasanen, Ilpo; Vuori, Erkki (2003-11-01). Toxicological screening with formula-based metabolite identification by liquid chromatography/time-of-flight mass spectrometry". Analytical Chemistry. **75** *(21): 5710–5718.* doi:10.1021/ac030162o. PMID 14588010.
26. Nobilis, Milan; Pour, Milan; Senel, Petr; Pavlík, Jan; Kunes, Jirí; Voprsalová, Marie; Kolárová, Lenka; Holcapek, Michal (2007-06-15). Metabolic profiling of a potential antifungal drug, 3-(4-bromophenyl)-5-acetoxymethyl-2,5-dihydrofuran-2-one, in mouse urine using high-performance liquid chromatography with UV photodiode-array and mass spectrometric detection". Journal of Chromatography B. **853** *(1–2): 10–19. doi:10.1016/j.jchromb.2007.02.045. PMID 17400036.*
27. Sundström, Mira; Pelander, Anna; Angerer, Verena; Hutter, Melanie; Kneisel, Stefan; Ojanperä, Ilkka (2013-10-01). A high-sensitivity ultra-high performance liquid chromatography/high-resolution time-of-flight mass spectrometry (UHPLC-HR-TOFMS) method for screening synthetic cannabinoids and other drugs of abuse in urine". Analytical and Bioanalytical Chemistry. **405** *(26): 8463–8474.* doi:10.1007/s00216-013-7272-8. PMID 23954996.
28. Zahedi Rad, Maliheh; Neyestani, Tirang Reza; Nikooyeh, Bahareh; Shariatzadeh, Nastaran; Kalayi, Ali; Khalaji, Niloufar; Gharavi, Azam (2015-01-01). Competitive Protein-binding assay-based Enzyme-immunoassay Method, Compared to High-pressure Liquid Chromatography, Has a Very Lower Diagnostic Value to Detect Vitamin D Deficiency in 9–12 Years Children". International Journal of Preventive Medicine. **6**: 67. doi:10.4103/2008-7802.161069. PMC 4542329. PMID 26330983.

CHAPTER 4

Gas chromatography–mass Spectrometry

Chapter Contents

- History
- Instrumentation
- Ionization
- Analysis
- Applications
- Liquid Chromatography–Mass Spectrometry
- Pyrolysis–Gas Chromatography–Mass Spectrometry
- Affinity Chromatography
- Weak Affinity Chromatography
- Advantages

A very powerful and sensitive instrument used to study trace amounts of chemicals in the air is a gas chromatograph (crow-MAT-oh-graf) connected to a mass spectrometer (spek-TRO-meh-ter), or GCMS. The GCMS can detect chemicals in amounts as small as a picogram. That is 0.000000000001 gram. One picogram is the equivalent of one drop of detergent in enough dishwater to fill a trainload of railroad tank cars ten miles long. Many of the pollutants found in air are present at concentrations lower than one picogram in a cubic meter of air. It is important for an the instrument to be able to detect these low concentrations.

Gas chromatography–mass spectrometry (**GC-MS**) is an analytical method that combines the features of gas-chromatography and mass spectrometry to identify different substances within a test sample. Applications of GC-MS include drug detection, fire investigation, environmental analysis, explosives investigation, and identification of unknown samples, including that of material samples obtained from planet Mars during probe missions as early as the 1970s. GC-MS can also be used in airport security to detect substances in luggage or on human beings. Additionally, it can identify trace elements in materials that were previously thought to have disintegrated beyond identification. Like liquid chromatography–mass spectrometry, it allows analysis and detection even of tiny amounts of a substance.

GC-MS has been widely heralded as a gold standard for forensic substance identification because it is used to perform a 100% specific test, which positively identifies the presence of a particular substance. A nonspecific test merely indicates that any of several in a category of substances is present. Although a nonspecific test could statistically suggest the identity of the substance, this could lead to false positive identification.

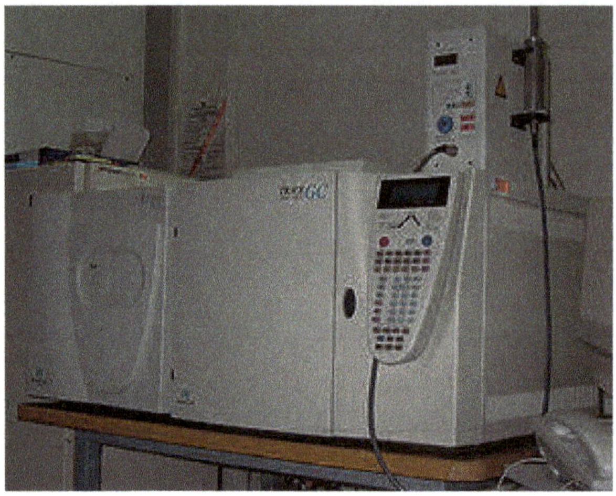

Example of a GC-MS instrument.

HISTORY

The use of a mass spectrometer as the detector in gas chromatography was developed during the 1950s after being originated by James and Martin in 1952. These comparatively sensitive devices were originally limited to laboratory settings.

The development of affordable and miniaturized computers has helped in the simplification of the use of this instrument, as well as allowed great improvements in the amount of time it takes to analyze a sample. In 1964, Electronic Associates, Inc. (EAI), a leading U.S. supplier of analog computers, began development of a computer controlled quadrupole mass spectrometer under the direction of Robert E. Finnigan. By 1966 Finnigan and collaborator Mike Uthe`s EAI division had sold over 500 quadrupole residual gas-analyzer instruments. In 1967, Finnigan left EAI to form the Finnigan Instrument Corporation along with Roger Sant, T. Z. Chou, Michael Story, and William Fies. In early 1968, they delivered the first prototype quadrupole GC/MS instruments to Stanford and Purdue University. When Finnigan Instrument Corporation was acquired by Thermo Instrument Systems (later Thermo Fisher Scientific) in 1990, it was considered the world`s leading manufacturer of mass spectrometers.

In 1996 the top-of-the-line high-speed GC-MS units completed analysis of fire accelerants in less than 90 seconds, whereas first-generation GC-MS would have required at least 16 minutes. By the 2000s computerized GC/MS instruments using quadrupole technology had become both essential to chemical research and one of the foremost instruments used for organic analysis. Today computerized GC/MS instruments are widely used in environmental monitoring of water, air, and soil; in the regulation of agriculture and food safety; and in the discovery and production of medicine.

INSTRUMENTATION

The insides of the GC-MS, with the column of the gas chromatograph in the oven on the right.

The GC-MS is composed of two major building blocks: the gas chromatograph and the mass spectrometer. The gas chromatograph utilizes a capillary column which depends on the column`s dimensions (length, diameter, film thickness) as well as the phase properties (e.g. 5% phenyl polysiloxane). The difference in the chemical properties between different molecules in a mixture and their relative affinity for the stationary phase of the column will promote separation of the molecules as the sample travels the length of the column. The molecules are retained by the column and then elute (come off) from the column at different times (called the retention time), and this allows the mass spectrometer downstream to capture, ionize, accelerate, deflect, and detect the ionized molecules separately. The mass spectrometer does this by breaking each molecule into ionized fragments and detecting these fragments using their mass-to-charge ratio.

These two components, used together, allow a much finer degree of substance identification than either unit used separately. It is not possible to make an accurate identification of a particular molecule by gas chromatography or mass spectrometry alone. The mass spectrometry process normally requires a very pure sample while gas chromatography using a traditional detector (e.g. Flame ionization detector) cannot differentiate between multiple molecules that happen to take the same amount of time to travel through the column (*i.e.* have the same retention time), which results in two or more molecules that co-elute. Sometimes two different molecules can also have a similar pattern of ionized fragments in a mass spectrometer (mass spectrum). Combining the two processes reduces the possibility of error, as it is extremely unlikely that two different molecules will behave in the same way in both a gas chromatograph and a mass spectrometer. Therefore, when an identifying mass spectrum appears at a characteristic retention time in a GC-MS analysis, it typically increases certainty that the analyte of interest is in the sample.

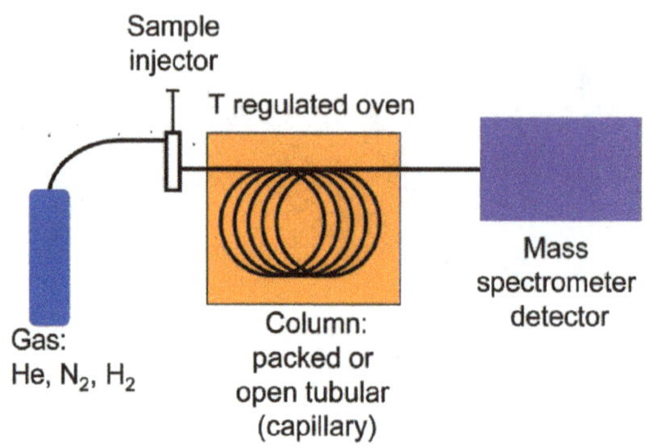

GC-MS schematic.

Purge and trap GC-MS

For the analysis of volatile compounds, a purge and trap (P&T) concentrator system may be used to introduce samples. The target analytes are extracted and mixed with water and introduced into an airtight chamber. An inert gas such as Nitrogen (N_2) is bubbled through the water; this is known as purging or sparging. The volatile compounds move into the headspace above the water and are drawn along a pressure gradient (caused by the introduction of the purge gas) out of the chamber. The volatile compounds are drawn along a heated line onto a 'trap'. The trap is a column of adsorbent material at ambient temperature that holds the compounds by returning them to the liquid phase. The trap is then heated and the sample compounds are introduced to the GC-MS column via a volatiles interface, which is a split inlet system. P&T GC-MS is particularly suited to volatile organic compounds (VOCs) and BTEX compounds (aromatic compounds associated with petroleum).

A faster alternative is the **"purge-closed loop"** system. In this system the inert gas is bubbled through the water until the concentrations of organic compounds in the vapor phase are at equilibrium with concentrations in the aqueous phase. The gas phase is then analysed directly.

Types of Mass Spectrometer Detectors

The most common type of mass spectrometer (MS) associated with a gas chromatograph (GC) is the quadrupole mass spectrometer, sometimes referred to by the Hewlett-Packard (now Agilent) trade name Mass Selective Detector (MSD). Another relatively common detector is the ion trap mass spectrometer. Additionally one may find a magnetic sector mass spectrometer, however these particular instruments are expensive and bulky and not typically found in high-throughput service laboratories. Other detectors may be encountered such as time of flight (TOF), tandem quadrupoles (MS-MS) (see below), or in the case of an ion trap MS^n where n indicates the number mass spectrometry stages.

GC-tandem MS

When a second phase of mass fragmentation is added, for example using a second quadrupole in a quadrupole instrument, it is called tandem MS (MS/MS). MS/MS can sometimes be used to quantitate low levels of target compounds in the presence of a high sample matrix background.

The first quadrupole (Q1) is connected with a collision cell (Q2) and another quadrupole (Q3). Both quadrupoles can be used in scanning or static mode, depending on the type of MS/MS analysis being performed. Types of analysis include product ion scan, precursor ion scan, selected reaction monitoring (SRM) (sometimes referred to as multiple reaction monitoring (MRM)) and neutral loss scan. For example: When Q1 is in static mode (looking at one mass only as in SIM), and Q3 is in scanning mode, one obtains a so-called product ion spectrum (also called daughter spectrum). From this spectrum, one can select a prominent product ion which can be the product ion for the chosen precursor ion. The pair is called a transition and forms the basis for SRM. SRM is highly specific and virtually eliminates matrix background.

IONIZATION

After the molecules travel the length of the column, pass through the transfer line and enter into the mass spectrometer they are ionized by various methods with typically only one method being used at any given time. Once the sample is fragmented it will then be detected, usually by an electron multiplier diode, which essentially turns the ionized mass fragment into an electrical signal that is then detected.

The ionization technique chosen is independent of using full scan or SIM.

Electron Ionization

By far the most common and perhaps standard form of ionization is electron ionization (EI). The molecules enter into the MS (the source is a quadrupole or the ion trap itself in an ion trap MS) where they are bombarded with free electrons emitted from a filament, not unlike the filament one would find in a standard light bulb. The electrons bombard the molecules, causing the molecule to fragment in a characteristic and reproducible way. This hard ionization technique results in the creation of more fragments of low mass to charge ratio (m/z) and few, if any, molecules approaching the molecular mass unit. Hard ionization is considered by mass spectrometrists as the employ of molecular electron bombardment, whereas soft ionization is charge by molecular collision with an introduced gas. The molecular fragmentation pattern is dependent upon the electron energy applied to the system, typically 70 eV (electron Volts). The use of 70 eV facilitates comparison of generated spectra with library spectra using manufacturer-supplied software or software developed by the National Institute of Standards (NIST-USA). Spectral library searches employ matching algorithms such as Probability Based Matching and dot-product matching that are used with methods of analysis written by many method standardization agencies. Sources of libraries include NIST, Wiley, the AAFS, and instrument manufacturers.

Cold electron Ionization

The "hard ionization" process of electron ionization can be softened by the cooling of the molecules before their ionization, resulting in mass spectra that are richer in information. In this method named cold electron ionization (Cold-EI) the molecules exit the GC column, mixed with added helium make up gas and expand into vacuum through a specially designed supersonic nozzle, forming a supersonic molecular beam (SMB). Collisions with the make up gas at the expanding supersonic jet reduce the internal vibrational (and rotational) energy of the analyte molecules, hence reducing the degree of fragmentation caused by the electrons during the ionization process. Cold-EI mass spectra are characterized by an abundant molecular ion while the usual fragmentation pattern is retained, thus making Cold-EI mass spectra compatible with library search identification techniques. The enhanced molecular ions increase the identification probabilities of both known and unknown compounds, amplify isomer mass spectral effects and enable the use of isotope abundance analysis for the elucidation of elemental formulae.

Chemical Ionization

In chemical ionization a reagent gas, typically methane or ammonia is introduced into the mass spectrometer. Depending on the technique (positive CI or negative CI) chosen, this reagent gas will interact with the electrons and analyte and cause a ‹soft` ionization of the molecule of interest. A softer ionization fragments the molecule to a lower degree than the hard ionization of EI. One of the main benefits of using chemical ionization is that a mass fragment closely corresponding to the molecular weight of the analyte of interest is produced.

In positive chemical ionization (PCI) the reagent gas interacts with the target molecule, most often with a proton exchange. This produces the species in relatively high amounts.

In negative chemical ionization (NCI) the reagent gas decreases the impact of the free electrons on the target analyte. This decreased energy typically leaves the fragment in great supply.

ANALYSIS

A mass spectrometer is typically utilized in one of two ways: full scan or selected ion monitoring (SIM). The typical GC-MS instrument is capable of performing both functions either individually or concomitantly, depending on the setup of the particular instrument.

The primary goal of instrument analysis is to quantify an amount of substance. This is done by comparing the relative concentrations among the atomic masses in the generated spectrum. Two kinds of analysis are possible, comparative and original. Comparative analysis essentially compares the given spectrum to a spectrum library to see if its characteristics are present for some sample in the library. This is best performed by a computer because there are a myriad of visual distortions that can take place due to variations in scale. Computers can also simultaneously correlate more data (such as the retention times identified by GC), to more accurately relate certain data.

Another method of analysis measures the peaks in relation to one another. In this method, the tallest peak is assigned 100% of the value, and the other peaks being assigned proportionate values. All values above 3% are assigned. The total mass of the unknown compound is normally indicated by the parent peak. The value of this parent peak can be used to fit with a chemical formula containing the various elements which are believed to be in the compound. The isotope pattern in the spectrum, which is unique for elements that have many isotopes, can also be used to identify the various elements present. Once a chemical formula has been matched to the spectrum, the molecular structure and bonding can be identified, and must be consistent with the characteristics recorded by GC-MS. Typically, this identification done automatically by programs which come with the instrument, given a list of the elements which could be present in the sample.

A "full spectrum" analysis considers all the "peaks" within a spectrum. Conversely, selective ion monitoring (SIM) only monitors selected ions associated with a specific substance. This is done on the assumption that at a given retention time, a set of ions is characteristic of a certain compound. This is a fast and efficient analysis, especially if the analyst has previous information about a sample or is only looking for a few specific substances. When the amount of information collected about the ions in a given gas chromatographic peak decreases, the sensitivity of the analysis increases. So, SIM analysis allows for a smaller quantity of a compound to be detected and measured, but the degree of certainty about the identity of that compound is reduced.

Full scan MS

When collecting data in the full scan mode, a target range of mass fragments is determined and put into the instrument's method. An example of a typical broad range of mass fragments to monitor would be *m/z* 50 to *m/z* 400. The determination of what range to use is largely dictated by what one anticipates being in the sample while being cognizant of the solvent and other possible interferences. A MS should not be set to look for mass fragments too low or else one may detect air (found as *m/z* 28 due to nitrogen), carbon dioxide (*m/z* 44) or other possible interference. Additionally if one is to use a large scan range then sensitivity of the instrument is decreased due to performing fewer scans per second since each scan will have to detect a wide range of mass fragments.

Full scan is useful in determining unknown compounds in a sample. It provides more information than SIM when it comes to confirming or resolving compounds in a sample. During instrument method development it may be common to first analyze test solutions in full scan mode to determine the retention time and the mass fragment fingerprint before moving to a SIM instrument method.

Selected ion Monitoring

In selected ion monitoring (SIM) certain ion fragments are entered into the instrument method and only those mass fragments are detected by the mass spectrometer. The advantages of SIM are that the detection limit is lower since the instrument is only looking at a small number of fragments (e.g. three fragments) during each scan. More scans can take place each second. Since only a few mass fragments of interest are being monitored, matrix interferences are typically lower. To additionally confirm the likelihood of a potentially positive result, it is relatively important to be sure that the ion ratios of the various mass fragments are comparable to a known reference standard.

APPLICATIONS

Environmental Monitoring and Cleanup

GC-MS is becoming the tool of choice for tracking organic pollutants in the environment. The cost of GC-MS equipment has decreased significantly, and the reliability has increased at the same time, which has contributed to its increased adoption in environmental studies.

Criminal Forensics

GC-MS can analyze the particles from a human body in order to help link a criminal to a crime. The analysis of fire debris using GC-MS is well established, and there is even an established American Society for Testing and Materials (ASTM) standard for fire debris analysis. GCMS/MS is especially useful here as samples often contain very complex matrices and results, used in court, need to be highly accurate.

Law Enforcement

GC-MS is increasingly used for detection of illegal narcotics, and may eventually supplant drug-sniffing dogs. It is also commonly used in forensic toxicology to find drugs and/or poisons in biological specimens of suspects, victims, or the deceased.

Sports anti-Doping Analysis

GC-MS is the main tool used in sports anti-doping laboratories to test athletes' urine samples for prohibited performance-enhancing drugs, for example anabolic steroids.

Security

A post–September 11 development, explosive detection systems have become a part of all US airports. These systems run on a host of technologies, many of them based on GC-MS. There are only three manufacturers certified by the FAA to provide these systems, one of which is Thermo Detection (formerly Thermedics), which produces the EGIS, a

GC-MS-based line of explosives detectors. The other two manufacturers are Barringer Technologies, now owned by Smith ‹s Detection Systems, and Ion Track Instruments, part of General Electric Infrastructure Security Systems.

Chemical Warfare Agent Detection

As part of the post-September 11 drive towards increased capability in homeland security and public health preparedness, traditional GC-MS units with transmission quadrupole mass spectrometers, as well as those with cylindrical ion trap (CIT-MS) and toroidal ion trap (T-ITMS) mass spectrometers have been modified for field portability and near real-time detection of chemical warfare agents (CWA) such as sarin, soman, and VX. These complex and large GC-MS systems have been modified and configured with resistively heated low thermal mass (LTM) gas chromatographs that reduce analysis time to less than ten percent of the time required in traditional laboratory systems. Additionally, the systems are smaller, and more mobile, including units that are mounted in mobile analytical laboratories (MAL), such as those used by the United States Marine Corps Chemical and Biological Incident Response Force MAL and other similar laboratories, and systems that are hand-carried by two-person teams or individuals, much ado to the smaller mass detectors. Depending on the system, the analytes can be introduced via liquid injection, desorbed from sorbent tubes through a thermal desorption process, or with solid-phase micro extraction (SPME).

Chemical Engineering

GC-MS is used for the analysis of unknown organic compound mixtures. One critical use of this technology is the use of GC-MS to determine the composition of bio-oils processed from raw biomass.

Food, beverage and perfume analysis

Foods and beverages contain numerous aromatic compounds, some naturally present in the raw materials and some forming during processing. GC-MS is extensively used for the analysis of these compounds which include esters, fatty acids, alcohols, aldehydes, terpenes etc. It is also used to detect and measure contaminants from spoilage or adulteration which may be harmful and which is often controlled by governmental agencies, for example pesticides.

Astrochemistry

Several GC-MS have left earth. Two were brought to Mars by the Viking program. Venera 11 and 12 and Pioneer Venus analysed the atmosphere of Venus with GC-MS. The Huygens probe of the Cassini-Huygens mission landed one GC-MS on Saturn`s largest moon, Titan. The material in the comet 67P/Churyumov-Gerasimenko will be analysed by the Rosetta mission with a chiral GC-MS in 2014.

Medicine

Dozens of congenital metabolic diseases also known as Inborn error of metabolism are now detectable by newborn screening tests, especially the testing using gas chromatography–mass spectrometry. GC-MS can determine compounds in urine even in minor concentration. These compounds are normally not present but appear in individuals suffering with metabolic disorders. This is increasingly becoming a common way to diagnose IEM for earlier diagnosis and institution of treatment eventually leading to a better outcome. It is now possible to test a newborn for over 100 genetic metabolic disorders by a urine test at birth based on GC-MS.

In combination with isotopic labeling of metabolic compounds, the GC-MS is used for determining metabolic activity. Most applications are based on the use of ^{13}C as the labeling and the measurement of ^{13}C-^{12}C ratios with an **isotope ratio mass spectrometer (IRMS)**; an MS with a detector designed to measure a few select ions and return values as ratios.

LIQUID CHROMATOGRAPHY–MASS SPECTROMETRY

Liquid chromatography–mass spectrometry (**LC-MS**, or alternatively **HPLC-MS**) is an analytical chemistry technique that combines the physical separation capabilities of liquid chromatography (or HPLC) with the mass analysis capabilities of mass spectrometry (MS). LC-MS is a powerful technique that has very high sensitivity, making it useful in many applications. Its application is oriented towards the separation, general detection and potential identification of chemicals of particular masses in the presence of other chemicals (i.e., in complex mixtures), e.g., natural products from natural-products extracts, and pure substances from mixtures of chemical intermediates. Preparative LC-MS systems can be used for rapid mass-directed purification of specific substances from such mixtures that are important in basic research, and pharmaceutical, agrochemical, food, and other industries. Like gas chromatography–mass spectrometry, it allows analysis and detection even of tiny amounts of a substance.

Liquid Chromatography

Present day liquid chromatography generally utilizes very small particles packed and operating at relatively high pressure, and is referred to as high performance liquid chromatography (HPLC); modern LC-MS methods use HPLC instrumentation, essentially exclusively, for sample introduction. In HPLC, the sample is forced by a liquid at high pressure (the mobile phase) through a column that is packed with a stationary phase generally composed of irregularly or spherically shaped particles chosen or derivatized to accomplish particular types of separations. HPLC methods are historically divided into two different sub-classes based on stationary phases and the corresponding required polarity of the mobile phase. Use of octadecylsilyl (C18) and related organic-modified particles as stationary phase with pure or pH-adjusted water-organic mixtures such as water-acetonitrile and water-methanol are used in techniques termed reversed phase liquid chromatography (RP-LC). Use of materials such as silica gel as stationary phase with neat or mixed organic mixtures are used in techniques termed normal phase liquid chromatography (NP-LC). RP-LC is most often used as the means to introduce samples into the MS, in LC-MS instrumentation.

Flow Splitting

When standard bore (4.6 mm) columns are used the flow is often split ~10:1. This can be beneficial by allowing the use of other techniques in tandem such as MS and UV detection. However splitting the flow to UV will decrease the sensitivity of spectrophotometric detectors. The mass spectrometry on the other hand will give improved sensitivity at flow rates of 200 μL/min or less.

Mass Spectrometry

Mass spectrometry (MS) is an analytical technique that measures the mass-to-charge ratio of charged particles. It is used for determining masses of particles, for determining the elemental composition of a sample or molecule, and for elucidating the chemical structures of molecules, such as peptides and other chemical compounds. MS works by ionizing chemical compounds to generate charged molecules or molecule fragments and measuring their mass-to-charge ratios. In a typical MS procedure, a sample is loaded onto the MS instrument and undergoes vaporization. The components of the sample are ionized by one of a variety of methods (e.g., by impacting them with an electron beam, using uv lights as a photon beam, laser beam or corona discharge etc.), which results in the formation of charged particles (ions). The ions are separated according to their mass-to-charge ratio in an analyzer by electromagnetic fields. The ions are detected, usually by a quantitative method. The ion signal is processed into mass spectra.

Additionally, MS instruments consist of three modules. An ion source, which can convert gas phase sample molecules into ions (or, in the case of electrospray ionization, move ions that exist in solution into the gas phase). A mass analyzer, which sorts the ions by their masses by applying electromagnetic fields. A detector, which measures the value of an indicator quantity and thus provides data for calculating the abundances of each ion present

The technique has both qualitative and quantitative uses. These include identifying unknown compounds, determining the isotopic composition of elements in a molecule, and determining the structure of a compound by observing its fragmentation. Other uses include quantifying the amount of a compound in a sample or studying the fundamentals of gas phase ion chemistry (the chemistry of ions and neutrals in a vacuum). MS is now in very common use in analytical laboratories that study physical, chemical, or biological properties of a great variety of compounds.

Mass Analyzer

There are many different mass analyzers that can be used in LC/MS. Single quadrupole, triple quadrupole, ion trap, time of flight (TOF) and quadrupole-time of flight (Q-TOF).

Interface

Understandably the interface between a liquid phase technique which continuously flows liquid, and a gas phase technique carried out in a vacuum was difficult for a long time. The advent of electrospray ionization changed this. The interface is most often an electrospray ion source or variant such as a nanospray source; however atmospheric pressure chemical ionization interface is also used. Various deposition and drying techniques have also been used such as using moving belts; however the most common of these is off-line MALDI deposition. A new approach still under development called Direct-EI LC-MS interface, couples a nano HPLC system and an electron ionization equipped mass spectrometer.

Applications

Pharmacokinetics

LC-MS is very commonly used in pharmacokinetic studies of pharmaceuticals and is thus the most frequently used technique in the field of bioanalysis. These studies give information about how quickly a drug will be cleared from the hepatic blood flow, and organs of the body. MS is used for this due to high sensitivity and exceptional specificity compared to UV (as long as the analyte can be suitably ionised), and short analysis time.

The major advantage MS has is the use of tandem MS-MS. The detector may be programmed to select certain ions to fragment. The process is essentially a selection technique, but is in fact more complex. The measured quantity is the sum of molecule fragments chosen by the operator. As long as there are no interferences or ion suppression, the LC separation can be quite quick.

Proteomics/Metabolomics

LC-MS is also used in proteomics where again components of a complex mixture must be detected and identified in some manner. The bottom-up proteomics LC-MS approach to proteomics generally involves protease digestion and denaturation (usually trypsin as a protease, urea to denature tertiary structure and iodoacetamide to cap cysteine residues) followed by LC-MS with peptide mass fingerprinting or LC-MS/MS (tandem MS) to derive sequence of individual peptides. LC-MS/MS is most commonly used for proteomic analysis of complex samples where peptide masses may overlap even with a high-resolution mass spectrometer. Samples of complex biological fluids like human serum may be run in a modern LC-MS/MS system and result in over 1000 proteins being identified, provided that the sample was first separated on an SDS-PAGE gel or HPLC-SCX.

Profiling of secondary metabolites in plants or food like phenolics can be achieved with liquid chromatography–mass spectrometry.

Drug Development

LC-MS is frequently used in drug development at many different stages including peptide mapping, glycoprotein mapping, natural products dereplication, bioaffinity screening, *in vivo* drug screening, metabolic stability screening, metabolite identification, impurity identification, quantitative bioanalysis, and quality control.

PYROLYSIS–GAS CHROMATOGRAPHY–MASS SPECTROMETRY

Pyrolysis–gas chromatography–mass spectrometry is a method of chemical analysis in which the sample is heated to decomposition to produce smaller molecules that are separated by gas chromatography and detected using mass spectrometry.

How it works

Pyrolysis is the thermal decomposition of materials in an inert atmosphere or a vacuum. The sample is put into direct contact with a platinum wire, or placed in a quartz sample tube, and rapidly heated to 600–1000 °C. Depending on the application even higher temperatures are used. Three different heating techniques are used in actual pyrolyzers: Isothermal furnace, inductive heating (Curie Point filament), and resistive heating using platinum filaments. Large molecules cleave at their weakest points and produce smaller, more volatile fragments. These fragments can be separated by gas chromatography. Pyrolysis GC chromatograms are typically complex because a wide range of different decomposition products is formed. The data can either be used as fingerprint to prove material identity or the GC/MS data is used to identify individual fragments to obtain structural information. To increase the volatility of polar fragments, various methylating reagents can be added to a sample before pyrolysis.

Besides the usage of dedicated pyrolyzers, pyrolysis GC of solid and liquid samples can be performed directly inside Programmable Temperature Vaporizer (PTV) injectors that provide quick heating (up to 60 °C/s) and high maximum temperatures of 600-650 °C. This is sufficient for many pyrolysis applications. The main advantage is that no dedicated instrument has to be purchased and pyrolysis can be performed as part of routine GC analysis. In this case quartz GC inlet liners can be used. Quantitative data can be acquired, and good results of derivatization inside the PTV injector are published as well.

Applications

Pyrolysis gas chromatography is useful for the identification of involatile compounds and synthetic polymeric media, such as acrylics or alkyds, and synthetic varnishes. The way in which the polymer fragments before it is separated in the GC can help in identifying the way that the sample polymer assembled itself (step growth v. chain growth, for example). The method of pyrolysis gas chromatography can also be used for environmental samples, including fossils. Trace evidence scientists in forensic laboratories use Pyroylsis GC to compare and identify paint or finish found at crime scenes or victims.

AFFINITY CHROMATOGRAPHY

Affinity chromatography is a method of separating biochemical mixtures based on a highly specific interaction such as that between antigen and antibody, enzyme and substrate, or receptor and ligand.

Uses

Affinity chromatography can be used to:

- Purify and concentrate a substance from a mixture into a buffering solution
- Reduce the amount of a substance in a mixture
- Discern what biological compounds bind to a particular substance
- Purify and concentrate an enzyme solution.

Principle

The stationary phase is typically a gel matrix, often of agarose; a linear sugar molecule derived from algae. Usually the starting point is an undefined heterogeneous group of molecules in solution, such as a cell lysate, growth medium or blood serum. The molecule of interest will have a well known and defined property, and can be exploited during the affinity purification process. The process itself can be thought of as an entrapment, with the target molecule becoming trapped on a solid or stationary phase or medium. The other molecules in the mobile phase will not become trapped as they do not possess this property. The stationary phase can then be removed from the mixture, washed and the target molecule released from the entrapment in a process known as elution. Possibly the most common use of affinity chromatography is for the purification of recombinant proteins.

Batch and Column Setups

Binding to the solid phase may be achieved by column chromatography whereby the solid medium is packed onto a column, the initial mixture run through the column to allow setting, a wash buffer run through the column and the elution buffer subsequently applied to the column and collected. These steps are usually done at ambient pressure. Alternatively, binding may be achieved using a batch treatment, for example, by adding the initial mixture to the solid phase in a vessel, mixing, separating the solid phase, removing the liquid phase, washing, re-centrifuging, adding the elution buffer, re-centrifuging and removing the elute.

Sometimes a hybrid method is employed such that the binding is done by the batch method, but the solid phase with the target molecule bound is packed onto a column and washing and elution are done on the column.

A third method, expanded bed absorption, which combines the advantages of the two methods mentioned above, has also been developed. The solid phase particles are placed in a column where liquid phase is pumped in from the bottom and exits at the top. The gravity of the particles ensure that the solid phase does not exit the column with the liquid phase.

Affinity columns can be eluted by changing salt concentrations, pH, pI, charge and ionic strength directly or through a gradient to resolve the particles of interest.

More recently, setups employing more than one column in series have been developed. The advantage compared to single column setups is that the resin material can be fully loaded, since non-binding product is directly passed on to a consecutive column with fresh column material. The resin costs per amount of produced product can thus be drastically be reduced. Since one column can always be eluted and regenerated while the other column is loaded, already two columns are sufficient to make full use of the advantages. Additional columns can give additional flexibility for elution and regeneration times, at the cost of additional equipment and resin costs.

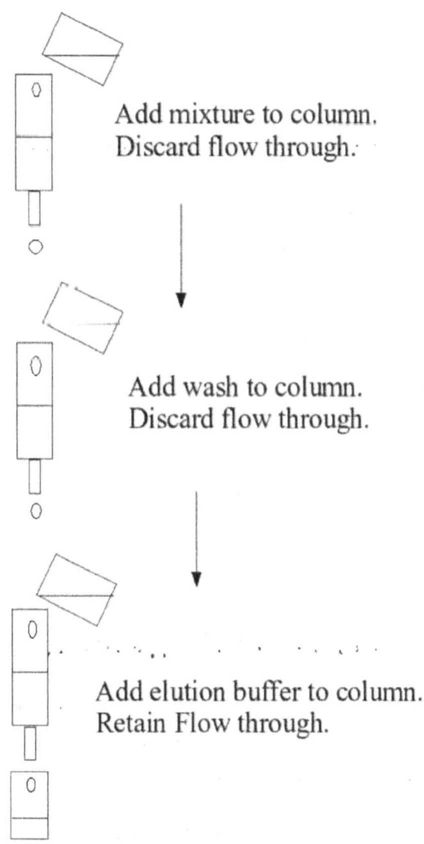

Column chromatography.

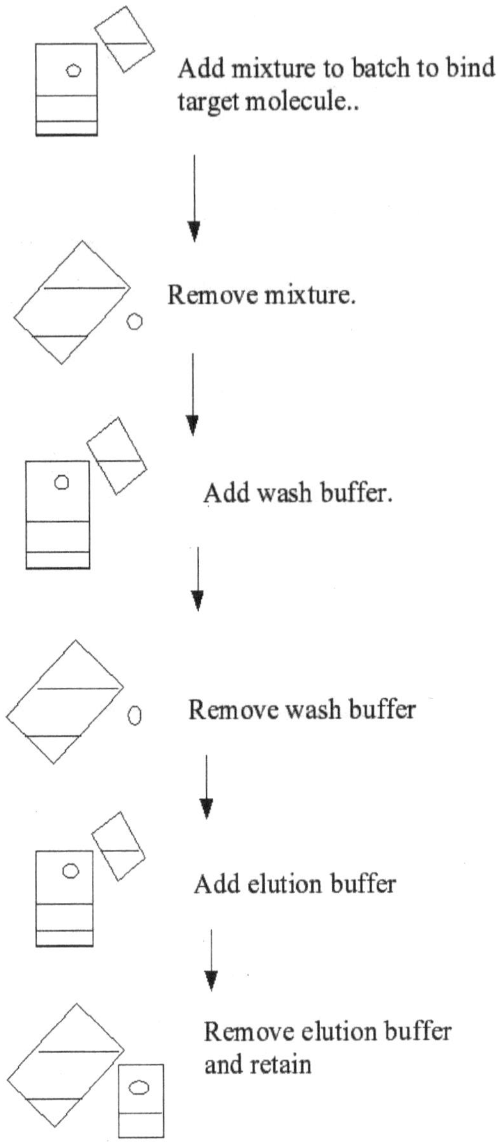

Batch chromatography.

Specific Uses

Affinity chromatography can be used in a number of applications, including nucleic acid purification, protein purification from cell free extracts, and purification from blood.

Various Affinity Media

Many different affinity media exist for a variety of possible uses. Briefly, they are (generalized):

- Activated/Functionalized – Works as a functional spacer, support matrix, and eliminates handling of toxic reagents.
- Amino Acid – Used with a variety of serum proteins, proteins, peptides, and enzymes, as well as rRNA and dsDNA.
- Avidin Biotin – Used in the purification process of biotin/avidin and their derivatives.
- Carbohydrate Bonding – Most often used with glycoproteins or any other carbohydrate-containing substance.
- Carbohydrate – Used with lectins, glycoproteins, or any other carbohydrate metabolite protein.
- Dye Ligand – This media is nonspecific, but mimics biological substrates and proteins.
- Glutathione – Useful for separation of GST tagged recombinant proteins.
- Heparin – This media is a generalized affinity ligand, and it is most useful for separation of plasma coagulation proteins, along with nucleic acid enzymes and lipases.
- Hydrophobic Interaction – Most commonly used to target free carboxyl groups and proteins.
- Immunoaffinity – Detailed below, this method utilizes antigens' and antibodies' high specificity to separate.
- Immobilized Metal Affinity Chromatography – Detailed further below, this method uses interactions between metal ions and proteins (usually specially tagged) to separate.
- Nucleotide/Coenzyme – Works to separate dehydrogenases, kinases, and transaminases.
- Nucleic Acid – Functions to trap mRNA, DNA, rRNA, and other nucleic acids/ oligonucleotides.
- Protein A/G – This method is used to purify immunoglobulins.
- Speciality – Designed for a specific class or type of protein/coenzyme, this type of media will only work to separate a specific protein or coenzyme.

Immunoaffinity

Another use for the procedure is the affinity purification of antibodies from blood serum. If serum is known to contain antibodies against a specific antigen (for example if the serum comes from an organism immunized against the antigen concerned) then it can be used for the affinity purification of that antigen. This is also known as Immunoaffinity Chromatography. For example, if an organism is immunised against a GST-fusion protein it will produce antibodies against the fusion-protein, and possibly antibodies against the GST tag as well. The protein can then be covalently coupled to a solid support such as agarose and used as an affinity ligand in purifications of antibody from immune serum.

For thoroughness the GST protein and the GST-fusion protein can each be coupled separately. The serum is initially allowed to bind to the GST affinity matrix. This will remove antibodies against the GST part of the fusion protein. The serum is then separated from the solid support and allowed to bind to the GST-fusion protein matrix. This allows any antibodies that recognize the antigen to be captured on the solid support. Elution of the antibodies of interest is most often achieved using a low pH buffer such as glycine pH 2.8. The eluate is collected into a neutral tris or phosphate buffer, to neutralize the low pH elution buffer and halt any degradation of the antibody`s activity. This is a nice example as affinity purification is used to purify the initial GST-fusion protein, to remove the undesirable anti-GST antibodies from the serum and to purify the target antibody.

A simplified strategy is often employed to purify antibodies generated against peptide antigens. When the peptide antigens are produced synthetically, a terminal cysteine residue is added at either the N- or C-terminus of the peptide. This cysteine residue contains a sulfhydryl functional group which allows the peptide to be easily conjugated to a carrier protein (e.g. Keyhole limpet hemocyanin (KLH)). The same cysteine-containing peptide is also immobilized onto an agarose resin through the cysteine residue and is then used to purify the antibody.

Most monoclonal antibodies have been purified using affinity chromatography based on immunoglobulin-specific Protein A or Protein G, derived from bacteria.

Immobilized metal ion Affinity Chromatography

Immobilized metal ion affinity chromatography (IMAC) is based on the specific coordinate covalent bond of amino acids, particularly histidine, to metals. This technique works by allowing proteins with an affinity for metal ions to be retained in a column containing immobilized metal ions, such as cobalt, nickel, copper for the purification of histidine containing proteins or peptides, iron, zinc or gallium for the purification of phosphorylated proteins or peptides. Many naturally occurring proteins do not have an affinity for metal ions, therefore recombinant DNA technology can be used to introduce such a protein tag into the relevant gene. Methods used to elute the protein of interest include changing the pH, or adding a competitive molecule, such as imidazole.

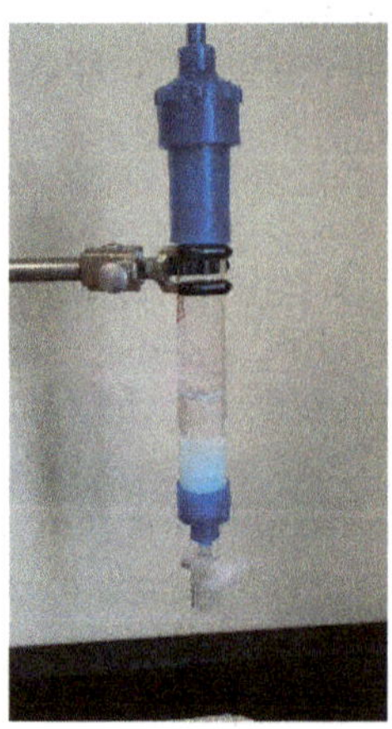

A chromatography column containing nickel-agarose beads used for purification of proteins with histidine tags

Recombinant Proteins

Possibly the most common use of affinity chromatography is for the purification of recombinant proteins. Proteins with a known affinity are protein tagged in order to aid their purification. The protein may have been genetically modified so as to allow it to be selected for affinity binding; this is known as a fusion protein. Tags include glutathione-S-transferase (GST), hexahistidine (His), and maltose binding protein (MBP). Histidine tags have an affinity for nickel or cobalt ions which have been immobilized by forming coordinate covalent bonds with a chelator incorporated in the stationary phase. For elution, an excess amount of a compound able to act as a metal ion ligand, such as imidazole, is used. GST has an affinity for glutathione which is commercially available immobilized as glutathione agarose. During elution, excess glutathione is used to displace the tagged protein.

Lectins

Lectin affinity chromatography is a form of affinity chromatography where lectins are used to separate components within the sample. Lectins, such as Concanavalin A are proteins which can bind specific carbohydrate (sugar) molecules. The most common application is to separate glycoproteins from non-glycosylated proteins, or one glycoform from another glycoform.

WEAK AFFINITY CHROMATOGRAPHY

Weak affinity chromatography (WAC) is a robust and highly reproducible technology for affinity screening in drug development. WAC is an affinity-based liquid chromatographic technique that separates chemical compounds based on their different weak affinities to an immobilized target.

The higher affinity a compound has towards the target, the longer it remains in the separation unit, and this will be expressed as a longer retention time. The affinity measure and ranking of affinity can be achieved by processing the obtained retention times of analyzed compounds.

Protein Targets

The WAC technology is demonstrated against a number of different protein targets – proteases, kinases, chaperones and protein–protein interaction (PPI) targets.

ADVANTAGES

WAC has been shown to be more effective than established methods for fragment based screening.

REFERENCES

1. James, A. T.; Martin, A. J. (1952). "Gas-liquid partition chromatography; the separation and micro-estimation of volatile fatty acids from formic acid to dodecanoic acid". The Biochemical Journal. 50 (5): 679–90. doi:10.1042/bj0500679. PMC 1197726. PMID 14934673.
2. Brock, David C. (2011). A Measure of Success. Chemical Heritage Magazine. 29 (1). Retrieved 22 April 2014.
3. Brock, David C. (2008). Detecting Success. Chemical Heritage Magazine. 26 (2): 31.
4. Thermo Instrument Systems Inc. History. International Directory of Company Histories (Volume 11 ed.). St. James Press. 1995. pp. 513–514. Retrieved 23 January 2015.
5. JeromeJeyakumar, J.; et. al. (April 2013). A Study of Phytochemical Constituents in Caralluma Umbellata By Gc – Ms Anaylsis [sic] (PDF). International Journal of Pharmaceutical Science Invention: 37–41. Retrieved 23 January 2015.
6. Optimizing the Analysis of Volatile Organic Compounds – Technical Guide Restek Corporation, Lit. Cat. 59887A
7. Wang, T.; Lenahan, R. (1984). Determination of volatile halocarbons in water by purge-closed loop gas chromatography. Bulletin of Environmental Contamination and Toxicology. 32 (1): 429–438. doi:10.1007/BF01607519. ISSN 0007-4861.
8. McLafferty, F. W.; Hertel, R. H.; Villwock, R. D. (1974). Probability based matching of mass spectra. Rapid identification of specific compounds in mixtures. Organic Mass Spectrometry. 9 (7): 690–702. doi:10.1002/oms.1210090710.
9. Stein, SE; Scott DR (1994). Optimization and testing of mass spectral library search algorithms for compound identification. J Am Soc Mass Spectrom. 5 (9): 859–866. doi:10.1016/1044-0305(94)87009-8. PMID 24222034.
10. Standard Reference Data. nist.gov
11. Wiley`s Scientific, Technical, and Medical Databases: Home. wiley.com
12. Mass Spectrometry Database Committee. ualberta.ca
13. Amirav A, Gordin A, Poliak M, Fialkov AB (2008). Gas chromatography-mass spectrometry with supersonic molecular beams. J Mass Spectrom. 43 (2): 141–63. doi:10.1002/jms.1380. PMID 18225851.
14. SMB-MS (Supersonic GC-MS). tau.ac.il
15. Alon, T.; Amirav, A. (2006). Isotope Abundance Analysis Method and Software for Improved Sample Identification with the Supersonic GC-MS. Rapid Communications in Mass Spectrometry. 20 (17): 2579–2588. doi:10.1002/rcm.2637. PMID 16897787.
16. Tsivou, M.; Kioukia-Fougia, N.; Lyris, E.; Aggelis, Y.; Fragkaki, A.; Kiousi, X.; Simitsek, P.; Dimopoulou, H.; Leontiou, I. -P.; Stamou, M.; Spyridaki, M. -H.; Georgakopoulos, C. (2006). An overview of the doping control analysis

during the Olympic Games of 2004 in Athens, Greece. Analytica Chimica Acta. 555: 1–13. doi:10.1016/j.aca.2005.08.068.

17. Smith, P. A.; Lepage, C. J.; Lukacs, M.; Martin, N.; Shufutinsky, A.; Savage, P. B. (2010). Field-portable gas chromatography with transmission quadrupole and cylindrical ion trap mass spectrometric detection: Chromatographic retention index data and ion/molecule interactions for chemical warfare agent identification. International Journal of Mass Spectrometry. 295 (3): 113–118. doi:10.1016/j.ijms.2010.03.001.

18. Sloan, K. M.; Mustacich, R. V.; Eckenrode, B. A. (2001). Development and evaluation of a low thermal mass gas chromatograph for rapid forensic GC-MS analyses. Field Analytical Chemistry & Technology. 5 (6): 288–301. doi:10.1002/fact.10011.

19. Patterson, G. E.; Guymon, A. J.; Riter, L. S.; Everly, M.; Griep-Raming, J.; Laughlin, B. C.; Ouyang, Z.; Cooks, R. G. (2002). Miniature Cylindrical Ion Trap Mass Spectrometer. Analytical Chemistry. 74 (24): 6145–6153. doi:10.1021/ac020494d. PMID 12510732.

20. Tekin, Kubilay; Karagöz, Selhan; Bektaş, Sema (2014-12-01). A review of hydrothermal biomass processing. Renewable and Sustainable Energy Reviews. 40: 673–687. doi:10.1016/j.rser.2014.07.216.

21. SEARCHING FOR LIFE ON MARS: The Development of the Viking GCMS. NASA

22. Krasnopolsky, V. A.; Parshev, V. A. (1981). Chemical composition of the atmosphere of Venus. Nature. 292 (5824): 610–613. doi:10.1038/292610a0.

23. Niemann, H. B.; Atreya, S. K.; Bauer, S. J.; Carignan, G. R.; Demick, J. E.; Frost, R. L.; Gautier, D.; Haberman, J. A.; Harpold, D. N.; Hunten, D. M.; Israel, G.; Lunine, J. I.; Kasprzak, W. T.; Owen, T. C.; Paulkovich, M.; Raulin, F.; Raaen, E.; Way, S. H. (2005). The abundances of constituents of Titan`s atmosphere from the GCMS instrument on the Huygens probe. Nature. 438 (7069): 779–784. doi:10.1038/nature04122. PMID 16319830.

24. Goesmann, F.; Rosenbauer, H.; Roll, R.; Böhnhardt, H. (2005). COSAC Onboard Rosetta: A Bioastronomy Experiment for the Short-Period Comet 67P/Churyumov-Gerasimenko. Astrobiology. 5 (5): 622–631. *Bibcode:2005AsBio...5..622G. doi:10.1089/ast.2005.5.622. PMID 16225435.*

CHAPTER 5

Capillary Electrochromatography

Chapter Contents

- **Introduction**
- **Principle**
- **Instrumentation and Working**
- **Advantages**

Capillary electrochromatography is an emerging analytical technique, which combines the versatility of chromatography with the high resolution of electrophoresis. The separation of complex analyte mixtures occurs as the result of a complex interplay between chromatographic retention and differential migration in the applied electric field. Such techniques use the electroosmotic flow (EOF) rather than a μ-pump to propulse the mobile phase, which reduces the number of mobile parts and renders the system more robust. The resolution of capillary electrochromatographic methods is extremely high. However, some problems prevent to date the use of the technique in routine analysis. Among them a lack of suitable algorithms to describe the separation. In addition the number of dedicated stationary phases and of proven applications remains small. In our work we focus in particular on stationary phase design and theoretical description of the separation mechanisms. In addition, we are very much interested in transferring the principle of capillary electrochromatography to microfluidic chips.

INTRODUCTION

Capillary electrochromatography (CEC) is a chromatographic technique in which the mobile phase is driven through the chromatographic bed by electroosmosis. Capillary electrochromatography is a combination of two analytical techniques, High performance liquid chromatography and capillary electrophoresis. Capillary electrophoresis aims to separate analytes on the basis of their mass to charge ratio by passing a high voltage across ends of a capillary tube, which is filled with the analyte. High performance liquid chromatography separates analytes by passing them, under high pressure, through a column filled with stationary phase. The interactions between the analytes and the stationary phase and mobile phase lead to the separation of the analytes. In capillary electrochromatography capillaries, packed with HPLC stationary phase, are subjected to a high voltage. Separation is achieved by electrophoretic migration of solutes and differential partitioning.

PRINCIPLE

Capillary electrochromatography (CEC) combines the principles used in HPLC and CE. The mobile phase is driven across the chromatographic bed using electroosmosis instead of pressure (as in HPLC). Electroosmosis is the motion of liquid induced by an applied potential across a porous material, capillary tube, membrane or any other fluid conduit. Electroosmotic flow is caused by the Coulomb force induced by an electric field on net mobile electric charge in a solution. Under alkaline conditions, the surface silanol groups of the fused silica will become ionised leading to a negatively charged surface. This surface will have a layer of positively charged ions in close proximity which are relatively immobilised. This layer of ions is called the Stern layer. The thickness of the double layer is given by the formula:

$$\delta = [\varepsilon r\ \varepsilon o\ \ RT / 2cF^2]^{1/2}$$

Where:

εr = dielectric constant or relative permittivity of medium

εo = permittivity of a vacuum

R = universal gas constant

T = absolute temperature

c = molar concentration

F = Faraday constant

When an electric field is applied to the fluid (usually via electrodes placed at inlets and outlets), the net charge in the electrical double layer is induced to move by the resulting Coulomb force. The resulting flow is termed electroosmotic flow. In CEC positive ions of the electrolyte added along with the analyte accumulate in the electrical double layer of the particles of the column packing on application of an electric field they move towards the cathode and drag the liquid mobile phase with them.

The relationship between the linear velocity u of the liquid in the capillary and the applied electric field is given by the Smoluchowski equation as follows:

$u = \varepsilon r \varepsilon o \zeta E \eta$

Where:

εr = Dielectric constant or relative permittivity of medium

εo = Permittivity of a vacuum

ζ = Potential across the Stern layer (Zeta Potential)

E = Electric field strength

η = Viscosity of the solvent

Separation of components in CEC is based on interactions between the stationary phase and differential electrophoretic migration of solutes.

INSTRUMENTATION AND WORKING

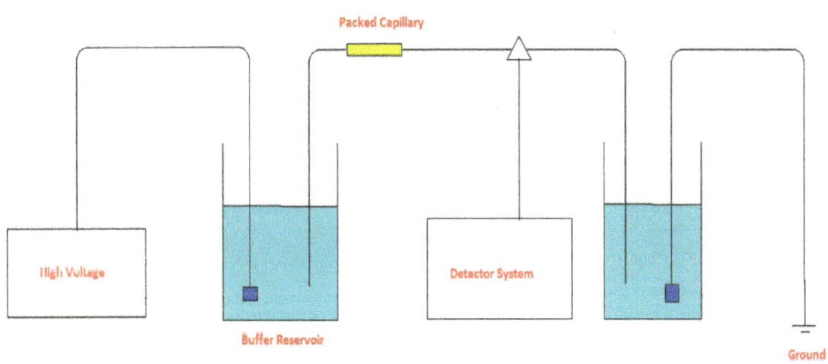

Figure 1: Diagram of Mechanism of Capillary Electrochromatography.

The basic components of an capillary electrochromatograph are a sample vial, source and destination vials, a packed capillary, electrodes, a high voltage power supply, a detector, and a data output and handling device. The source vial, destination vial and capillary are filled with an electrolyte such as an aqueous buffer solution. The capillary is packed with stationary phase. To introduce the sample, the capillary inlet is placed into a vial containing the sample and then returned to the source vial (sample is introduced into the capillary via capillary action, pressure, or siphoning). The migration of the analytes is then initiated by an electric field that is applied between the source and destination vials and is supplied to the electrodes by the high-voltage power supply. The analytes separate as they migrate due to their electrophoretic mobility, and are detected near the outlet end of the capillary. The output of the detector is sent to a data output and handling device such as an integrator or computer. The data is then displayed as an electropherogram, which reports detector response as a function of time. Separated chemical compounds appear as peaks with different migration times in an electropherogram.

ADVANTAGES

Avoiding the use of pressure to introduce the mobile phase into the column, results in a number of important advantages. Firstly, the pressure driven flow rate across a column depends directly on the square of the particle diameter and inversely on the length of the column. This restricts the length of the column and size of the particle, particle size is seldom less than 3 micrometer and the length of the column is restricted to 25 cm. Electrically driven flow rate is independent of length of column and size .A second advantage of using electroosmosis to pass the mobile phase into the column is the plug like flow velocity profile of EOF reduces the solute dispersion in the column, increasing column efficiency.

REFERENCES

1. Smith, N. “Capillary ElectroChromatography” Available at:https://www.beckmancoulter.com/wsrportal/bibliography?docname=AP8508ACECPrimer.pdf
2. Bartle, K. D. *Capillary ElectroChromatography* Published by The Royal Society of Chemistry, Cambridge. ISBN 0-85404-530-9

CHAPTER 6

Separation Mechanism Techniques

Chapter Contents

- Ion Chromatography
- History
- Principle
- Procedure
- Weak and Strong Ion Exchangers
- Typical Technique
- Membranc Exchange Chromatography
- Separating Proteins
- Uses
- Size-Exclusion Chromatography
- Expanded Bed Adsorption

The separation mechanism depends upon differences in polarity between the different feed components. The more polar a molecule, the more strongly it will be adsorbed by a polar stationary phase. Similarly, the more non-polar a molecule, the more strongly it will be adsorbed by non-polar stationary phase.

During a surface adsorption chromatography process, there is competition for stationary phase adsorption sites, between the materials to be separated and the mobile phase. Feed molecules of low polarity spend proportionally more time in the mobile phase than those molecules that are highly polar, which are retained longer. Therefore the components of a mixture are eluted in order of increasing polarity.

ION CHROMATOGRAPHY

Ion chromatography (or **ion-exchange chromatography**) is a chromatography process that separates ions and polar molecules based on their affinity to the ion exchanger. It works on almost any kind of charged molecule—including large proteins, small nucleotides, 0061nd amino acids. It is often used in protein purification, water analysis, and quality control. The water-soluble and charged molecules such as proteins, amino acids, and peptides bind to moieties which are oppositely charged by forming ionic bonds to the insoluble stationary phase. The equilibrated stationary phase consists of an ionizable functional group where the targeted molecules of a mixture to be separated and quantified can bind while passing through the column—a cationic stationary phase is used to separate anions and an anionic stationary phase is used to separate cations. Cation exchange chromatography is used when the desired molecules to separate are cations and anion exchange chromatography is used to separate anions. The bound molecules then can be eluted and collected using an eluant which contains anions and cations by running higher concentration of ions through the column or changing pH of the column. One of the primary advantages for the use of ion chromatography is only one interaction involved during the separation as opposed to other separation techniques; therefore, ion chromatography may have higher matrix tolerance. However, there are also disadvantages involved when performing ion-exchange chromatography, such as constant evolution with the technique which leads to the inconsistency from column to column.

Dionex ICS-3000 ion chromatography system.

HISTORY

Table: History of ion exchange and ion chromatography, the analytical technique based on ion exchange

ca. 1850	Soil as an ion exchanger for Mg2+, Ca2+ and NH4+	Thomson & Way
1935	Sulfonated and aminated condensation polymers (phenol/formaldehyde)	Adams, Homes
1942	Sulfonated PS/DVB resin as cation exchanger (Manhattan Project)	d'Alelio
1947	Aminated PS/DVB resin as anion exchanger	McBurney
1953	Ion exclusion chromatography	Wheaton, Baumann
1957	Macroporous ion exchangers	Corte, Meyer, Kunin et al.
1959	Basic Theoretical principles	Helfferich
1967-70	Pellicular ion exchnagers	Horvath, Kirkland
1975	Ion exchange chromatography with conductivity detection using a "stripper"	Small, Stevens, Bauann
1979	Conductivity detection without a "stripper"	Gjerde, Fritz, Schmuckler
1976-80	Ion pair chromatography	Waters, Bidlingmeier, Horvath et al.

Ion exchange chromatography.

Ion chromatography has advanced through the accumulation of knowledge over a course of many years. Starting from 1947, Spedding and Powell used displacement ion-exchange chromatography for the separation of the rare earths. Additionally, they showed the ion-exchange separation of 14N and 15N isotopes in ammonia. At the start of the 1950s, Kraus and Nelson demonstrated the use of many analytical methods for metal ions dependent on their separation of their chloride, fluoride, nitrate or sulfate complexes by anion chromatography. Automatic in-line detection was progressively introduced from 1960 to 1980 as well as novel chromatographic methods for metal ion separations. A groundbreaking method by Small, Stevens and Bauman at Dow Chemical Co. unfolded the creation of the modern ion chromatography. Anions and cations could now be separated efficiently by a system of suppressed conductivity detection. In 1979, a method for anion chromatography with non-suppressed conductivity detection was introduced by Gjerde et al. Following it in 1980, was a similar method for cation chromatography.

As a result, a period of extreme competition began within the IC market, with supporters for both suppressed and nonsuppressed conductivity detection. This competition led to fast growth of new forms and the fast evolution of IC. A challenge that needs to be overcome in the future development of IC is the preparation of highly efficient monolithic ion-exchange columns and overcoming this challenge would be of great importance to the development of IC.

The boom of Ion exchange chromatography primarily began between 1935–1950 during World War II and it was through the "Manhattan project" that applications and IC were significantly extended. Ion chromatography was originally introduced by two English researchers, agricultural Sir Thompson and chemist J T Way. The works of Thompson and Way involved the action of water-soluble fertilizer salts, ammonium sulfate and potassium chloride. These salts could not easily be extracted from the ground due to the rain. They performed ion methods to treat clays with the salts, resulting in the extraction of ammonia in addition to the release of calcium. It was in the fifties and sixties that theoretical models were developed for IC for further understanding and it was not until the seventies that continuous detectors were utilized, paving the path for the development from low-pressure to high-performance chromatography. Not until 1975 was "ion chromatography" established as a name in reference to the techniques, and was thereafter used as a name for marketing purposes. Today IC is important for investigating aqueous systems, such as drinking water. It is a popular method for analyzing anionic elements or complexes that help solve environmentally relevant problems. Likewise, it also has great uses in the semiconductor industry.

Because of the abundant separating columns, elution systems, and detectors available, chromatography has developed into the main method for ion analysis.

When this technique was initially developed, it was primarily used for water treatment. Since 1935, ion exchange chromatography rapidly manifested into one of the most heavily leveraged techniques, with its principles often being applied to majority of fields of chemistry, including distillation, adsorption, and filtration.

PRINCIPLE

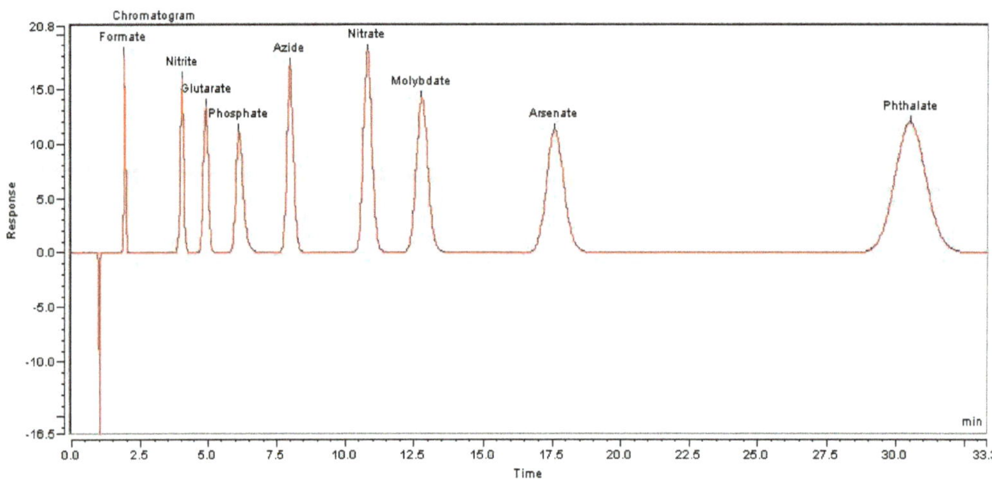

Ion Chromatography.

Ion-exchange chromatography separates molecules based on their respective charged groups. Ion-exchange chromatography retains analyte molecules on the column based on coulombic (ionic) interactions. Essentially, molecules undergo electrostatic interactions with opposite charges on the stationary phase matrix. The stationary phase consists of an immobile matrix that contains charged ionizable functional groups or ligands. The stationary phase surface displays ionic functional groups (R-X) that interact with analyte ions of opposite charge. To achieve electroneutrality, these inert charges couple with exchangeable counterions in the solution. Ionizable molecules that are to be purified compete with these exchangeable counterions for binding to the immobilized charges on the stationary phase. These ionizable molecules are retained or eluted based on their charge. Initially, molecules that do not bind or bind weakly to the stationary phase are first to wash away. Altered conditions are needed for the elution of the molecules that bind to the stationary phase. The concentration of the exchangeable counterions, which competes with the molecules for binding, can be increased or the pH can be changed. A change in pH affects the charge on the particular molecules and, therefore, alters binding. The molecules then start eluting out based on the changes in their charges from the adjustments. Further such adjustments can be used to release the protein of interest. Additionally, concentration of counterions can be gradually varied to separate ionized molecules. This type of elution is called gradient elution. On the other hand, step elution can be used in which the concentration of counterions are varied in one step. This type of chromatography is further subdivided into cation exchange chromatography and anion-exchange chromatography. Positively charged molecules bind to anion exchange resins while negatively charged molecules bind to cation exchange resins. The ionic compound consisting of the cationic species M+ and the anionic species B- can be retained by the stationary phase.

Cation exchange chromatography retains positively charged cations because the stationary phase displays a negatively charged functional group:

$$R\text{-}X^{-}C^{+} + M^{+}B^{-} \rightleftarrows R\text{-}X^{-}M^{+} + C^{+} + B^{-}$$

Anion exchange chromatography retains anions using positively charged functional group:

$$R\text{-}X^{+}A^{-} + M^{+}B^{-} \rightleftarrows R\text{-}X^{+}B^{-} + M^{+} + A^{-}$$

Note that the ion strength of either C+ or A- in the mobile phase can be adjusted to shift the equilibrium position, thus retention time.

The ion chromatogram shows a typical chromatogram obtained with an anion exchange column.

PROCEDURE

It is possible to perform ion exchange chromatography in bulk, on thin layers of medium such as glass or plastic plates coated with a layer of the desired stationary phase, or in chromatography columns. Thin layer chromatography or column chromatography share similarities in that they both act within the same governing principles; there is constant and frequent exchange of molecules as the mobile phase travels along the stationary phase. It is not imperative to add the sample in minute volumes as the predetermined conditions for the exchange column have been chosen so that there will be strong interaction between the mobile and stationary phases. Furthermore, the mechanism of the elution process will cause a compartmentalization of the differing molecules based on their respective chemical characteristics. This phenomenon is due to an increase in salt concentrations at or near the top of the column, thereby displacing the molecules at that position, while molecules bound lower are released at a later point when the higher salt concentration reaches that area. These principles are the reasons that ion exchange chromatography is an excellent candidate for initial chromatography steps in a complex purification procedure as it can quickly yield small volumes of target molecules regardless of a greater starting volume.

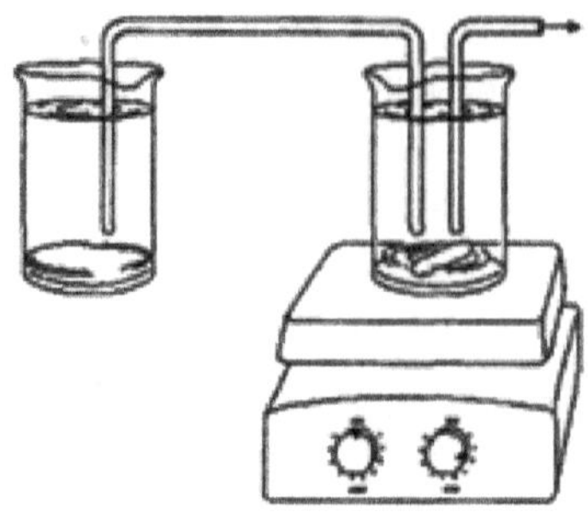

Chamber (left) contains high salt concentration. Stirred chamber (right) contains low salt concentration. Gradual stirring causes the formation of a salt gradient as salt travel from high to low concentrations.

Comparatively simple devices are often used to apply counterions of increasing gradient to a chromatography column. Counterions such as Copper(II) are chosen most often for effectively separating peptides and amino acids through complex formation.

A simple device can be used to create a salt gradient. Elution buffer is consistently being drawn from the chamber into the mixing chamber, thereby altering its buffer concentration. Generally, the buffer placed into the chamber is usually of high initial concentration, whereas the buffer placed into the stirred chamber is usually of low concentration. As the high concentration buffer from the left chamber is mixed and drawn into the column, the buffer concentration of the stirred column gradually increase. Altering the shapes of the stirred chamber, as well as of the limit buffer, allows for the production of concave, linear, or convex gradients of counterion.

A multitude of different mediums are used for the stationary phase. Among the most common immobilized charged groups used are trimethylaminoethyl (TAM), triethylaminoethyl (TEAE), diethyl-2-hydroxypropylaminoethyl (QAE), aminoethyl (AE), diethylaminoethyl (DEAE), sulpho (S), sulphomethyl (SM), sulphopropyl (SP), carboxy (C), and carboxymethyl (CM).

Successful column packing is an important aspect of ion chromatography. Stability and efficiency of a final column depends on packing methods, solvent used, and factors that affect mechanical properties of the column. In contrast to early inefficient dry- packing methods, wet slurry packing, in which particles that are suspended in an appropriate solvent are delivered into a column under pressure, shows significant improvement. Three different approaches can be employed in performing wet slurry packing: the balanced density method (solvent's density is about that of porous silica particles), the high viscosity method (a solvent of high viscosity is used), and the low viscosity slurry method (performed with low viscosity solvents).

Polystyrene is used as a medium for ion- exchange. It is made from the polymerization of styrene with the use of divinylbenzene and benzoyl peroxide. Such exchangers form hydrophobic interactions with proteins which can be irreversible. Due to this property, polystyrene ion exchangers are not suitable for protein separation. They are used on the other hand for the separation of small molecules in amino acid separation and removal of salt from water. Polystyrene ion exchangers with large pores can be used for the separation of protein but must be coated with a hydrophillic substance.

Cellulose based medium can be used for the separation of large molecules as they contain large pores. Protein binding in this medium is high and has low hydrophobic character. DEAE is an anion exchange matrix that is produced from a positive side group of diethylaminoethyl bound to cellulose or Sephadex.

Agarose gel based medium contain large pores as well but their substitution ability is lower in comparison to dextrans. The ability of the medium to swell in liquid is based on the cross-linking of these substances, the pH and the ion concentrations of the buffers used.

Incorporation of high temperature and pressure allows a significant increase in the efficiency of ion chromatography, along with a decrease in time. Temperature has an influence of selectivity due to its effects on retention properties. The retention factor ($k = (t_R^g - t_M^g)/(t_M^g - t_{ext})$) increases with temperature for small ions, and the opposite trend is observed for larger ions.

Despite ion selectivity in different mediums, further research is being done to perform ion exchange chromatography through the range of 40–175 °C.

An appropriate solvent can be chosen based on observations of how column particles behave in a solvent. Using an optical microscope, one can easily distinguish a desirable dispersed state of slurry from aggregated particles.

WEAK AND STRONG ION EXCHANGERS

A "strong" ion exchanger will not lose the charge on its matrix once the column is equilibrated and so a wide range of pH buffers can be used. "Weak" ion exchangers have a range of pH values in which they will maintain their charge. If the pH of the buffer used for a weak ion exchange column goes out of the capacity range of the matrix, the column will lose its charge distribution and the molecule of interest may be lost. Despite the smaller pH range of weak ion exchangers, they are often used over strong ion exchangers due to their having greater specificity. In some experiments, the retention times of weak ion exchangers are just long enough to obtain desired data at a high specificity.

Resins of ion exchange columns may include functional groups such as weak/strong acids and weak/strong bases. There are also special columns that have resins with amphoteric functional groups that can exchange both cations and anions.Examples of functional groups of strong ion exchange resins are quaternary ammonium (Q), which is an anion exchanger, and sulfonic acid (S), which is a cation exchanger. These types of exchangers can maintain their charge density over a pH range of 0–14. Examples of functional groups of Weak ion exchange resins include diethylaminoethyl (DEAE), which is an anion exchanger, and carboxymethyl (CM), which is a cation exchanger. These two types of exchangers can maintain the charge density of their columns over a pH range of 5–9.

In ion chromatography, the interaction of the solute ions and the stationary phase based on their charges determines which ions will bind and to what degree. When the stationary phase features positive groups which attracts anions, it is called an anion exchanger; when there are negative groups on the stationary phase, cations are attracted and it is a cation exchanger. The attraction between ions and stationary phase also depends on the resin, organic particles used as ion exchangers.

Each resin features relative selectivity which varies based on the solute ions present who will compete to bind to the resin group on the stationary phase. The selectivity coefficient, the equivalent to the equilibrium constant, is determined via a ratio of the concentrations between the resin and each ion, however, the general trend is that ion exchangers prefer binding to the ion with a higher charge, smaller hydrated radius, and higher polarizability, or the ability for the electron cloud of an ion to be disrupted by other charges. Despite this selectivity, excess amounts of an ion with a lower selectivity introduced to the column would cause the lesser ion to bind more to the stationary phase as the selectivity coefficient allows fluctuations in the binding reaction that takes place during ion exchange chromatography.

TYPICAL TECHNIQUE

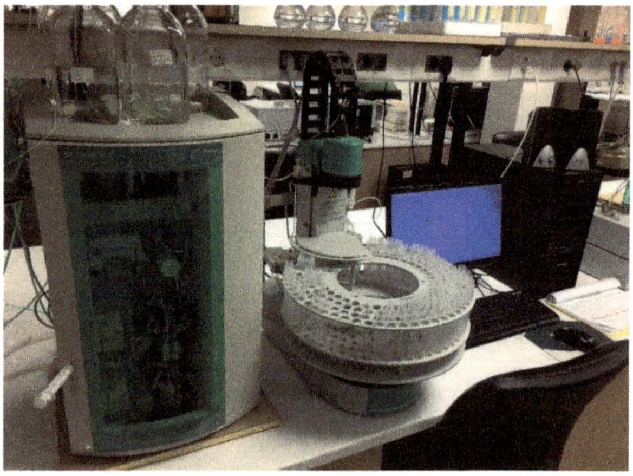

Metrohm 850 Ion chromatography system.

A sample is introduced, either manually or with an autosampler, into a sample loop of known volume. A buffered aqueous solution known as the mobile phase carries the sample from the loop onto a column that contains some form of stationary phase material. This is typically a resin or gel matrix consisting of agarose or cellulose beads with covalently bonded charged functional groups. Equilibration of the stationary phase is needed in order to obtain the desired charge of the column. If the column is not properly equilibrated the desired molecule may not bind strongly to the column. The target analytes (anions or cations) are retained on the stationary phase but can be eluted by increasing the concentration of a similarly charged species that displaces the analyte ions from the stationary phase. For example, in cation exchange chromatography, the positively charged analyte can be displaced by adding positively charged sodium ions. The analytes of interest must then be detected by some means, typically by conductivity or UV/visible light absorbance.

Control an IC system usually requires a chromatography data system (CDS). In addition to IC systems, some of these CDSs can also control gas chromatography (GC) and HPLC.

MEMBRANE EXCHANGE CHROMATOGRAPHY

A type of ion exchange chromatography, membrane exchange is a relatively new method of purification designed to overcome limitations of using columns packed with beads. Membrane Chromatographic devices are cheap to mass-produce and disposable unlike other chromatography devices that require maintenance and time to revalidate. There are three types of membrane absorbers that are typically used when separating substances. The three types are flat sheet, hollow fibre, and radial flow. The most common absorber and best suited for membrane chromatography is multiple flat sheets because it has more absorbent volume. It can be used to overcome mass transfer limitations and pressure drop, making it especially advantageous for isolating and purifying viruses, plasmid DNA, and other large macromolecules. The column is packed with microporous membranes with internal pores which contain adsorptive moieties that can bind the target protein. Adsorptive membranes are available in a variety of geometries and chemistry which allows them to be used for purification and also fractionation, concentration, and clarification in an efficiency that is 10 fold that of using beads. Membranes can be prepared through isolation of the membrane itself, where membranes are cut into squares and immobilized. A more recent method involved the use of live cells that are attached to a support membrane and are used for identification and clarification of signaling molecules.

SEPARATING PROTEINS

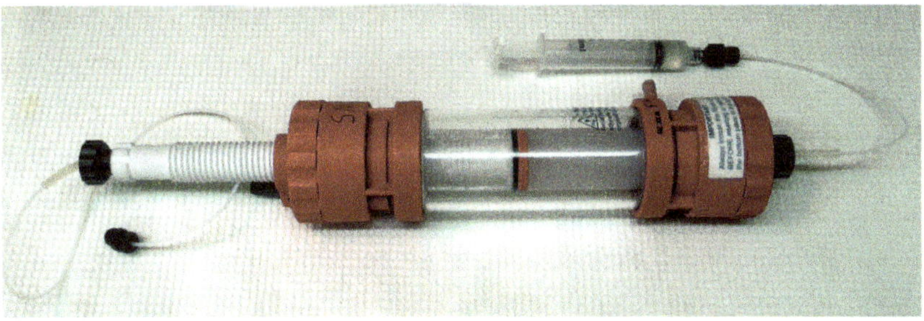

Preparative-scale ion exchange column used for protein purification.

In ion exchange chromatography can be used to separate proteins because proteins are composed of charged functional groups. The ions of interest (in this case charged proteins) are exchanged for another ion, (which is usually salt) on a charged solid support. Usually the solutes are in a liquid phase, which tends to be water. Take for example proteins in water, which would be a liquid phase that is passed through a column. The column is commonly known as the solid phase since it is filled with porous synthetic particles that are of a particular charge. These porous particles are also referred to as beads, may be aminated (containing amino groups) or have metal ions in order to have a charge. The column can be prepared using porous polymers, for macromolecules over 100,000 the optimum size of the porous particle is about 1 μm^2. This is because slow diffusion of the solutes within the pores does not restrict the separation quality. The beads containing positively charged groups are attracted to the negatively charged protein, commonly referred to as anion exchange resins. Proteins that have negatively at pH 7 (pH of water) charged side chains are glutamate and aspartate. Beads that are negatively charged will be cation exchange resins, as positively charged protein will attracted. The proteins that have positively charged side chains at pH 7 are lysine, histidine and asparagine.

The isoelectric point is the pH at which a compound in this case a protein has no net charge. A protein's isoelectric point or PI can be determines using the pka of the side chains, if the amino (positive chain) is able to cancel out the carbonyl (negative) chain, the protein would be at its PI. Using buffers instead of water for proteins that do not have a charge at pH 7, is a good idea as it enables the manipulation of pH to alter ionic interactions between the proteins and the beads. Weakly acid or basic proteins (such as leucine, proline, alanine, valine, glycine...etc.) are able to have a charge of the pH is high enough to deprotonate the amino group. Separation can be achieved based on the natural isoelectric point of the protein. Alternatively a peptide tag can be genetically added to the protein to give the protein an isoelectric point away from most natural proteins (e.g., 6 arginines for binding to a cation-exchange resin or 6 glutamates for binding to an anion-exchange resin such as DEAE-Sepharose).

Elution by increasing ionic strength of the mobile phase is more subtle. It works because ions from the mobile phase interact with the immobilized ions on the stationary phase, thus "shielding" the stationary phase from the protein, and letting the protein elute.

Elution from ion-exchange columns can be sensitive to changes of a single charge-chromatofocusing. Ion-exchange chromatography is also useful in the isolation of specific multimeric protein assemblies, allowing purification of specific complexes according to both the number and the position of charged peptide tags.

Gibbs-Donnan Effect

In ion exchange chromatography, the Gibbs–Donnan effect is observed when the pH of the applied buffer and the ion exchanger differ, even up to one pH unit. For example, in anion-exchange columns, the ion exchangers repeal protons so the pH of the buffer near the column differs is higher than the rest of the solvent. As a result, an experimenter has to be careful that the protein(s) of interest is stable and properly charged in the "actual" pH.

This effect comes as a result of two similarly charged particles, one from the resin and one from the solution, failing to distribute properly between the two sides; there is a selective uptake of one ion over another. For example, in a sulphonated polystyrene resin, a cation exchange resin, the chlorine ion of a hydrochloric acid buffer should equilibrate into the resin. However, since the concentration of the sulphonic acid in the resin is high, the hydrogen of HCl has no tendency to enter the column. This, combined with the need of electroneutrality, leads to a minimum amount of hydrogen and chlorine entering the resin.

USES

Clinical Utility

A use of ion chromatography can be seen in the argentation ion chromatography. Usually silver and compounds containing acetylenic and ethylenic bonds have very weak interactions. This phenomenon has been widely tested on olefin compounds. The ion complexes the olefins make with silver ions are weak and made based on the overlapping of pi, sigma, and d orbitals and available electrons therefore cause no real changes in the double bond. This behavior was manipulated to separate lipids, mainly fatty acids from mixtures in to fractions with differing number of double bonds using silver ions. The ion resins were impregnated with silver ions, which were then exposed to various acids (silicic acid) to elute fatty acids of different characteristics.

Detection limits as low as 1 μM can be obtained for alkali metal ions. It may be used for measurement of HbA1c, porphyrin and with water purification. Ion Exchange Resins(IER) have been widely used especially in medicines due to its high capacity and the uncomplicated system of the separation process. One of the synthetic uses is to use Ion Exchange Resins for kidney dialysis. This method is used to separate the blood elements by using the cellulose membraned artificial kidney.

Another clinical application of ion chromatography is in the rapid anion exchange chromatography technique used to separate creatine kinase (CK) isoenzymes from human serum and tissue sourced in autopsy material (mostly CK rich tissues were used such as cardiac muscle and brain). These isoenzymes include MM, MB, and BB, which all carry out the same function given different amino acid sequences. The functions of these isoenzymes are to convert creatine, using ATP, into phosphocreatine expelling ADP. Mini columns were filled with DEAE-Sephadex A-50 and further eluted with tris- buffer sodium chloride at various concentrations (each concentration was chosen advantageously to manipulate elution). Human tissue extract was inserted in columns for separation. All fractions were analyzed to see total CK activity and it was found that each source of CK isoenzymes had characteristic isoenzymes found within. Firstly, CK- MM was eluted, then CK-MB, followed by CK-BB. Therefore, the isoenzymes found in each sample could be used to identify the source, as they were tissue specific.

Using the information from results, correlation could be made about the diagnosis of patients and the kind of CK isoenzymes found in most abundant activity. From the finding, about 35 out of 71 patients studied suffered from heart attack (myocardial infarction) also contained an abundant amount of the CK-MM and CK-MB isoenzymes. Findings further show that many other diagnosis including renal failure, cerebrovascular disease, and pulmonary disease were only found to have the CK-MM isoenzyme and no other isoenzyme. The results from this study indicate correlations between various diseases and the CK isoenzymes found which confirms previous test results using various techniques. Studies about CK-MB found in heart attack victims have expanded since this study and application of ion chromatography.

Industrial Applications

Since 1975 ion chromatography has been widely used in many branches of industry. The main beneficial advantages are reliability, very good accuracy and precision, high selectivity, high speed, high separation efficiency, and low cost of consumables. The most significant development related to ion chromatography are new sample preparation methods; improving the speed and selectivity of analytes separation; lowering of limits of detection and limits of quantification; extending the scope of applications; development of new standard methods; miniaturization and extending the scope of the analysis of a new group of substances. Allows for quantitative testing of electrolyte and proprietary additives of electroplating baths. It is an advancement of qualitative hull cell testing or less accurate UV testing. Ions, catalysts, brighteners and accelerators can be measured. Ion exchange chromatography has gradually become a widely known, universal technique for the detection of both anionic and cationic species. Applications for such purposes have been developed, or are under development, for a variety of fields of interest, and in particular, the pharmaceutical industry. The usage of ion exchange chromatography in pharmaceuticals has increased in recent years, and in 2006, a chapter on ion exchange chromatography was officially added to the United States Pharmacopia-National Formulary (USP-NF). Furthermore, in 2009 release of the USP-NF, the United States Pharmacopia made several analyses of ion chromatography available using two techniques: conductivity detection, as well as pulse amperometric detection. Majority of these applications are primarily used for measuring and analyzing residual limits in pharmaceuticals, including detecting the limits of oxalate, iodide, sulfate, sulfamate, phosphate, as well as various electrolytes including potassium, and sodium. In total, the 2009 edition of the USP-NF officially released twenty eight methods of detection for the analysis of active compounds, or components of active compounds, using either conductivity detection or pulse amperometric detection.

Drug Development

There has been a growing interest in the application of IC in the analysis of pharmaceutical drugs. IC is used in different aspects of product development and quality control testing. For example, IC is used to improve stabilities and solubility properties of pharmaceutical active drugs molecules as well as used to detect systems that have higher tolerance for organic solvents. IC has been used for the determination of analytes as a part of a dissolution test. For instance, calcium dissolution tests have shown that other ions present in the medium can be well resolved among themselves and also from the calcium ion. Therefore, IC has been employed in drugs in the form of tablets and capsules in order to determine the amount of drug dissolve with time. IC is also widely used for detection and quantification of excipients or inactive ingredients used in pharmaceutical formulations. Detection of sugar and sugar alcohol in such formulations through IC has been done due to these polar groups getting resolved in ion column. IC methodology also established in analysis of impurities in drug substances and products. Impurities or any components that are not part of the drug chemical entity are evaluated and they give insights about the maximum and minimum amounts of drug that should be administered in a patient per day.

SIZE-EXCLUSION CHROMATOGRAPHY

Size-exclusion chromatography (SEC), also known as **molecular sieve chromatography**, is a chromatographic method in which molecules in solution are separated by their size, and in some cases molecular weight. It is usually applied to large molecules or macromolecular complexes such as proteins and industrial polymers. Typically, when an aqueous solution is used to transport the sample through the column, the technique is known as **gel-filtration chromatography**, versus the name **gel permeation chromatography**, which is used when an organic solvent is used as a mobile phase. SEC is a widely used polymer characterization method because of its ability to provide good molar mass distribution (Mw) results for polymers.

Applications

The main application of gel-filtration chromatography is the fractionation of proteins and other water-soluble polymers, while gel permeation chromatography is used to analyze the molecular weight distribution of organic-soluble polymers. Either technique should not be confused with gel electrophoresis, where an electric field is used to pull or push molecules through the gel depending on their electrical charges.

Advantages

The advantages of this method include good separation of large molecules from the small molecules with a minimal volume of eluate, and that various solutions can be applied without interfering with the filtration process, all while preserving the biological activity of the particles to be separated. The technique is generally combined with others that further separate molecules by other characteristics, such as acidity, basicity, charge, and affinity for certain compounds. With size exclusion chromatography, there are short and well-defined separation times and narrow bands, which lead to good sensitivity. There is also no sample loss because solutes do not interact with the stationary phase.

The other advantage to this experimental method is that in certain cases it is feasible to determine the approximate molecular weight of a compound. The shape and size of the compound (eluent) determine how the compound interacts with the gel (stationary phase). In order to determine the approximate molecular weight, the elution volumes of compounds with their corresponding molecular weights are obtained and then a plot of "K_{av}" vs "log(Mw)" is made, where $K_{av} = (V_e-V_o)/(V_t-V_o)$ and Mw is the molecular mass. This plot acts as a calibration curve, which is used to approximate the desired compound's molecular weight. The V_e component represents the volume at which the intermediate molecules elute such as molecules that have partial access to the beads of the column. In addition, V_t is the sum of the total volume between the beads and the volume within the beads. The V_o component represents the volume at which the larger molecules elute, which elute in the beginning.Disadvantages are, for example, that only a limited number of bands can be accommodated because the time scale of the chromatogram is short, and, in general, there has to be a 10% difference in molecular mass to have a good resolution

Discovery

The technique was invented by Grant Henry Lathe and Colin R Ruthven, working at Queen Charlotte's Hospital, London. They later received the John Scott Award for this invention. While Lathe and Ruthven used starch gels as the matrix, Jerker Porath and Per Flodin later introduced dextran gels;other gels with size fractionation properties include agarose and polyacrylamide. A short review of these developments has appeared.

There were also attempts to fractionate synthetic high polymers; however, it was not until 1964, when J. C. Moore of the Dow Chemical Company published his work on the preparation of gel permeation chromatography (GPC) columns based on cross-linked polystyrene with controlled pore size, that a rapid increase of research activity in this field began. It was recognized almost immediately that with proper calibration, GPC was capable to provide molar mass and molar mass distribution information for synthetic polymers. Because the latter information was difficult to obtain by other methods, GPC came rapidly into extensive use.

Theory and Method

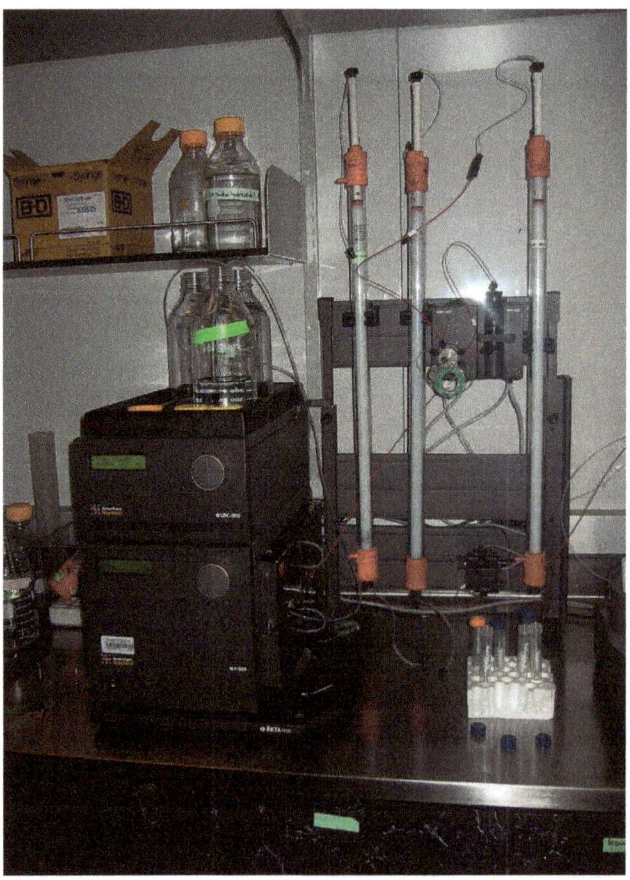

Agarose-based SEC columns used for protein purification on an AKTA FPLC machine.

SEC is used primarily for the analysis of large molecules such as proteins or polymers. SEC works by trapping smaller molecules in the pores of the adsorbent materials adsorption ("stationary phases"). This process is usually performed with a column, which consists of a hollow tube tightly packed with extremely small porous polymer beads designed to have pores of different sizes. These pores may be depressions on the surface or channels through the bead. As the solution travels down the column some particles enter into the pores. Larger particles cannot enter into as many pores. The larger the particles, the faster the elution. The larger molecules simply pass by the pores because those molecules are too large to enter the pores. Larger molecules therefore flow through the column more quickly than smaller molecules, that is, the smaller the molecule, the longer the retention time.

One requirement for SEC is that the analyte does not interact with the surface of the stationary phases, with differences in elution time between analytes ideally being based solely on the solute volume the analytes can enter, rather than chemical or electrostatic interactions with the stationary phases. Thus, a small molecule that can penetrate every region of the stationary phase pore system can enter a total volume equal to the sum of the entire pore volume and the interparticle volume. This small molecule will elute late (after the molecule has penetrated all of the pore- and interparticle volume—approximately 80% of the column volume). At the other extreme, a very large molecule that cannot penetrate any the smaller pores can enter only the interparticle volume (~35% of the column volume) and will elute earlier when this volume of mobile phase has passed through the column. The underlying principle of SEC is that particles of different sizes will elute (filter) through a stationary phase at different rates. This results in the separation of a solution of particles based on size. Provided that all the particles are loaded simultaneously or near-simultaneously, particles of the same size should elute together.

However, as there are various measures of the size of a macromolecule (for instance, the radius of gyration and the hydrodynamic radius), a fundamental problem in the theory of SEC has been the choice of a proper molecular size parameter by which molecules of different kinds are separated. Experimentally, Benoit and co-workers found an excellent correlation between elution volume and a dynamically based molecular size, the hydrodynamic volume, for several different chain architecture and chemical compositions. The observed correlation based on the hydrodynamic volume became accepted as the basis of universal SEC calibration.

Still, the use of the hydrodynamic volume, a size based on dynamical properties, in the interpretation of SEC data is not fully understood. This is because SEC is typically run under low flow rate conditions where hydrodynamic factor should have little effect on the separation. In fact, both theory and computer simulations assume a thermodynamic separation principle: the separation process is determined by the equilibrium distribution (partitioning) of solute macromolecules between two phases --- a dilute bulk solution phase located at the interstitial space and confined solution phases within the pores of column packing material. Based on this theory, it has been shown that the relevant size parameter to the partitioning of polymers in pores is the mean span dimension (mean maximal projection onto a line).Although this issue has not been fully resolved, it is likely that the mean span dimension and the hydrodynamic volume are strongly correlated.

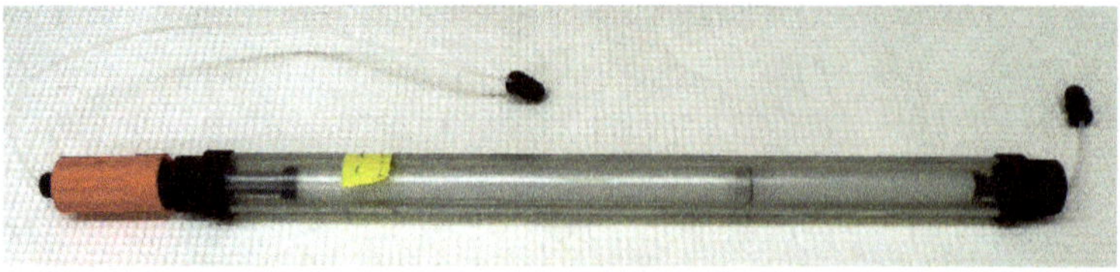

A size exclusion column.

Each size exclusion column has a range of molecular weights that can be separated. The exclusion limit defines the molecular weight at the upper end of the column 'working' range and is where molecules are too large to be trapped in the stationary phase. The lower end of the range is defined by the permeation limit, which defines the molecular weight of a molecule that is small enough to penetrate all pores of the stationary phase. All molecules below this molecular mass are so small that they elute as a single band

The filtered solution that is collected at the end is known as the **eluate**. The **void volume** includes any particles too large to enter the medium, and the solvent volume is known as the **column volume**.

Factors Affecting Filtration

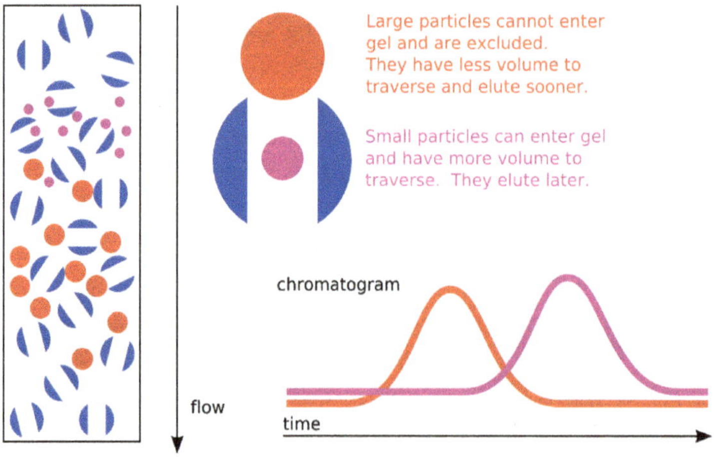

A cartoon illustrating the theory behind **size exclusion chromatography.**

In real-life situations, particles in solution do not have a fixed size, resulting in the probability that a particle that would otherwise be hampered by a pore passing right by it. Also, the stationary-phase particles are not ideally defined; both particles and pores may vary in size. Elution curves, therefore, resemble Gaussian distributions. The stationary phase may also interact in undesirable ways with a particle and influence retention times, though great care is taken by column manufacturers to use stationary phases that are inert and minimize this issue.

Like other forms of chromatography, increasing the column length will enhance the resolution, and increasing the column diameter increases the capacity of the column. Proper column packing is important to maximize resolution: An overpacked column can collapse the pores in the beads, resulting in a loss of resolution. An underpacked column can reduce the relative surface area of the stationary phase accessible to smaller species, resulting in those species spending less time trapped in pores. Unlike affinity chromatography techniques, a solvent head at the top of the column can drastically diminish resolution as the sample diffuses prior to loading, broadening the downstream elution.

Analysis

In simple manual columns, the eluent is collected in constant volumes, known as fractions. The more similar the particles are in size the more likely they will be in the same fraction and not detected separately. More advanced columns overcome this problem by constantly monitoring the eluent.

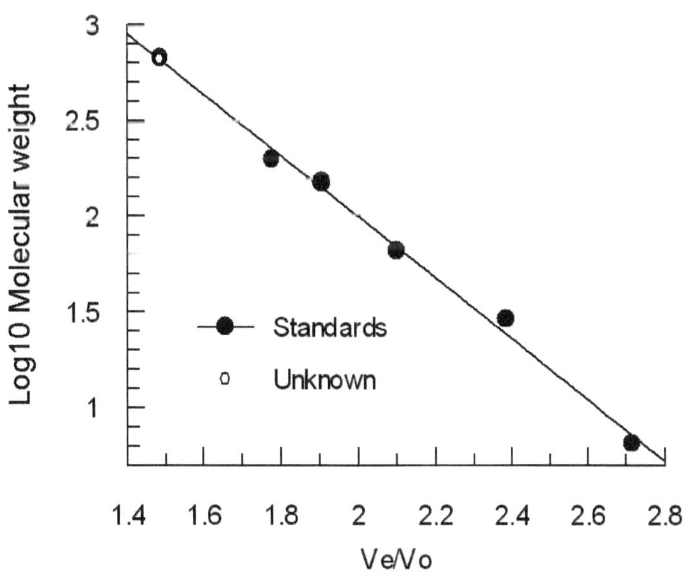

Standardization of a size exclusion column.

The collected fractions are often examined by spectroscopic techniques to determine the concentration of the particles eluted. Common spectroscopy detection techniques are refractive index (RI) and ultraviolet (UV). When eluting spectroscopically similar species (such as during biological purification), other techniques may be necessary to identify the contents of each fraction. It is also possible to analyse the eluent flow continuously with RI, LALLS, Multi-Angle Laser Light Scattering MALS, UV, and/or viscosity measurements.

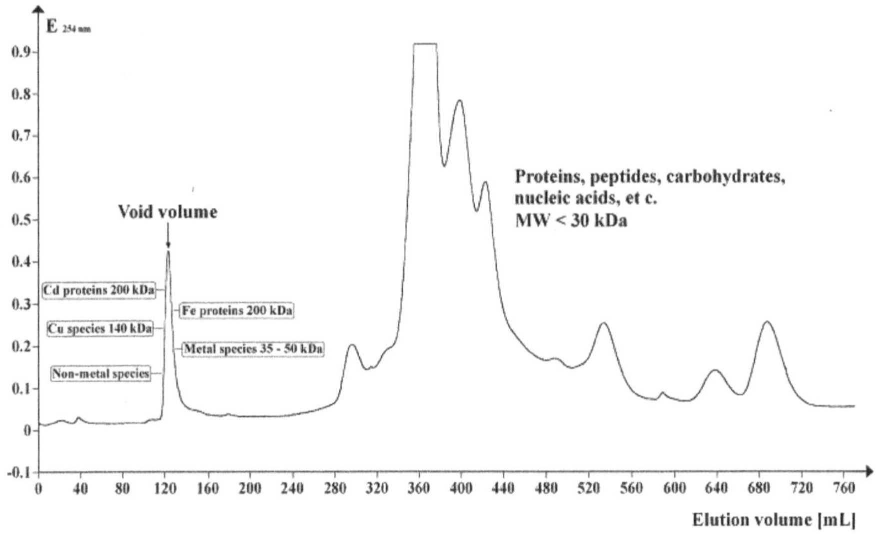

SEC Chromatogram of a **biological sample**.

The elution volume (Ve) decreases roughly linear with the logarithm of the molecular hydrodynamic volume. Columns are often calibrated using 4-5 standard samples (e.g., folded proteins of known molecular weight), and a sample containing a very large molecule such as thyroglobulin to determine the void volume. (Blue dextran is not recommended for Vo determination because it is heterogeneous and may give variable results) The elution volumes of the standards are divided by the elution volume of the thyroglobulin (Ve/Vo) and plotted against the log of the standards` molecular weights.

Applications

Biochemical Applications

In general, SEC is considered a low resolution chromatography as it does not discern similar species very well, and is therefore often reserved for the final "polishing" step of a purification. The technique can determine the quaternary structure of purified proteins that have slow exchange times, since it can be carried out under native solution conditions, preserving macromolecular interactions. SEC can also assay protein tertiary structure, as it measures the hydrodynamic volume (not molecular weight), allowing folded and unfolded versions of the same protein to be distinguished. For example, the apparent hydrodynamic radius of a typical protein domain might be 14 Å and 36 Å for the folded and unfolded forms, respectively. SEC allows the separation of these two forms, as the folded form will elute much later due to its smaller size.\

Polymer Synthesis

SEC can be used as a measure of both the size and the polydispersity of a synthesised polymer, that is, the ability to be able to find the distribution of the sizes of polymer molecules. If standards of a known size are run previously, then a calibration curve can be created to determine the sizes of polymer molecules of interest in the solvent chosen for analysis (often THF). In alternative fashion, techniques such as light scattering and/or viscometry can be used online with SEC to yield absolute molecular weights that do not rely on calibration with standards of known molecular weight. Due to the difference in size of two polymers with identical molecular weights, the absolute determination methods are, in general, more desirable. A typical SEC system can quickly (in about half an hour) give polymer chemists information on the size and polydispersity of the sample. The preparative SEC can be used for polymer fractionation on an analytical scale.

Drawback

In SEC, mass is not measured so much as the hydrodynamic volume of the polymer molecules, that is, how much space a particular polymer molecule takes up when it is in solution. However, the approximate molecular weight can be calculated from SEC data because the exact relationship between molecular weight and hydrodynamic volume for polystyrene can be found. For this, polystyrene is used as a standard. But the relationship between hydrodynamic volume and molecular weight is not the same for all polymers, so only an approximate measurement can be obtained. Another drawback is the possibility of interaction between the stationary phase and the analyte. Any interaction leads to a later elution time and thus mimics a smaller analyte size.

When performing this method, the bands of the eluting molecules may be broadened. This can occur by turbulence caused by the flow of the mobile phase molecules passing through the molecules of the stationary phase. In addition, molecular thermal diffusion and friction between the molecules of the glass walls and the molecules of the eluent contribute to the broadening of the bands. Besides broadening, the bands also overlap with each other. As a result, the eluent usually gets considerably diluted. A few precautions can be taken to prevent the likelihood of the bands broadening. For instance, one can apply the sample in a narrow, highly concentrated band on the top of the column. The more concentrated the eluent is, the more efficient the procedure would be. However, it is not always possible to concentrate the eluent, which can be considered as one more disadvantage.

Absolute Size-Exclusion Chromatography

Absolute size-exclusion chromatography (ASEC) is a technique that couples a dynamic light scattering (DLS) instrument to a size exclusion chromatography system for absolute size measurements of proteins and macromolecules as they elute from the chromatography system.

The definition of absolute used here is that it does not require calibration to obtain hydrodynamic size, often referred to as hydrodynamic diameter (D_H in units of nm). The sizes of the macromolecules are measured as they elute into the flow cell of the DLS instrument from the size exclusion column set. It should be noted that the hydrodynamic size of the molecules or particles are measured and not their molecular weights. For proteins a Mark-Houwink type of calculation can be used to estimate the molecular weight from the hydrodynamic size.

A big advantage of DLS coupled with SEC is the ability to obtain enhanced DLS resolution. Batch DLS is quick and simple and provides a direct measure of the average size but the baseline resolution of DLS is 3 to 1 in diameter. Using SEC, the proteins and protein oligomers are separated, allowing oligomeric resolution. Aggregation studies can also be done using ASEC although the aggregate concentration may not be calculated, the size of the aggregate will be measured only to be limited by the maximum size eluting from the SEC columns.

Limitations of ASEC include flow-rate, concentration, and precision. Because a correlation function requires anywhere from 3–7 seconds to properly build, a limited number of data points can be collected across the peak.

EXPANDED BED ADSORPTION

Expanded bed adsorption (EBA) is a preparative chromatographic technique which makes processing of viscous and particulate liquids possible.

Principle

The protein binding principles in EBA are the same as in classical column chromatography and the common ion-exchange, hydrophobic interaction and affinity chromatography ligands can be used. After the adsorption step is complete, the fluidized bed is washed to flush out any remaining particulates. Elution of the adsorbed proteins was commonly performed with the eluent flow in the reverse direction; that is, as a conventional packed bed, in order to recover the adsorbed solutes in a smaller volume of eluent. However, a new generation of EBA columns has been developed, which maintain the bed in the expanded state during this phase, producing high-purity, high yields of e.g. MAbs [monoclonal antibodies] in even smaller volumes of eluent. Process duration at manufacturing scale has also been cut considerably (under 7 hours in some cases).

EBA may be considered to combine both the Removal of Insolubles and the Isolation steps of the 4-step downstream processing heuristic. The major limitations associated with EBA technology is biomass interactions and aggregations onto adsorbent during processing.

Where classical column chromatography uses a solid phase made by a packed bed, EBA uses particles in a fluidized state, ideally expanded by a factor of 2. Expanded bed adsorption is, however, different from fluidised bed chromatography in essentially two ways: one, the EBA resin contains particles of varying size and density which results in a gradient of particle size when expanded; and two, when the bed is in its expanded state, local loops are formed. Particles such as whole cells or cell debris, which would clog a packed bed column, readily pass through a fluidized bed. EBA can therefore be used on crude culture broths or slurries of broken cells, thereby bypassing initial clearing steps such as centrifugation and filtration, which is mandatory when packed beds are used. In older EBA column designs, the feed flow rate is kept low enough that the solid packing remains stratified and does not fluidize completely. Hence EBA can be modelled as frontal adsorption in a packed bed, rather than as a well-mixed, continuous-flow adsorber.

REFERENCES

1. Luqman, Mohammad, and Inamuddin (2012). Ion Exchange Technology II. Springer Netherlands. p. 1. ISBN 978-94-007-4026-6.
2. Fritz, James S. (1987). Ion chromatography. Analytical Chemistry. **59** *(4): 335A.* doi:10.1021/ac00131a002.
3. Neubauer, Kenneth. Advantages and Disadvantages of Different Column Types for Speciation Analysis by LC-ICP-MS. spectroscopyonline.com. Retrieved 9 May 2016.
4. Fritz, J. S. (2004). Early milestones in the development of ion-exchange chromatography: a personal account. Journal of chromatography. A. **1039** *(1–2): 3–12.* doi:10.1016/s0021-9673(04)00020-2. PMID 15250395.
5. Lucy, C. A. (2003). Evolution of ion-exchange: from Moses to the Manhattan Project to Modern Times. Journal of chromatography. A. **1000** *(1–2): 711–24.* doi:10.1016/s0021-9673(03)00528-4. PMID 12877196.
6. Recent developments and emerging directions in ion chromatography.
7. Andylong. The History Of Ion Exchange Chromatography. Ion Chromatography. Retrieved 9 May 2016.
8. Eith, Claudia, Kolb Maximilian, and Seubert Andreas (2002). Introduction to Practical Ion Chromatography An Introduction. Ed. Viehweger Kai. Herisau: Metrohm. p. 160.
9. Nachod, F. C. (1940). Ion Exchange: Theory and Application. Academic Press. pp. 1, 2. ISBN 0124312993.
10. Jungbauer, Alois; Hahn, Rainer (2009). Chapter 22 Ion-Exchange Chromatography. Guide to Protein Purification, 2nd Edition. Methods in Enzymology. **463**. pp. 349–371. doi:10.1016/S0076-6879(09)63022-6. ISBN 9780123745361.
11. ^ Jump up to:[a] [b] Ninfa, Alexander J., David P.Ballou, and Marilee Benore (2010). Fundamental Laboratory Approaches for Biochemistry and Biotechnology. Hoboken, NJ: John Wiley
12. Duong-Ly, K. C.; Gabelli, S. B. (2014). Using Ion Exchange Chromatography to Purify a Recombinantly Expressed Protein. Laboratory Methods in Enzymology: Protein Part C. Methods in enzymology. Methods in Enzymology. **541**. pp. 95–103. doi:10.1016/B978-0-12-420119-4.00008-2. ISBN 9780124201194. PMID 24674065.
13. Ninfa, Alexander; Ballou, David; Benore, Marilee (May 26, 2009). Fundamental Laboratory Approaches for Biochemistry and Biotechnology. Wiley. pp. 143–145. ISBN 0470087668.

14. Dieter, Debra S.; Walton, Harold F. (1983-11-01). Counterion effects in ion-exchange partition chromatography. Analytical Chemistry. **55** *(13): 2109–2112.* doi:10.1021/ac00263a025.
15. Kirkland, J. J.; DeStefano, J. J. (2006-09-08). The art and science of forming packed analytical high-performance liquid chromatography columns. Journal of Chromatography A. The Role of Theory in Chromatography. **1126** *(1–2): 50–57. doi:10.1016/j.chroma.2006.04.027.*
16. Janson, Jan-Christer. (2011). Protein Purification: Principles, High Resolution Methods, and Applications. John Wiley & Sons.
17. Lackie, John (2010). A Dictionary of Biochemistry. Oxford University Press. ISBN 9780199549351.
18. Wouters, Bert; Bruggink, Cees; Desmet, Gert; Agroskin, Yury; Pohl, Christopher A.; Eeltink, Sebastiaan (2012-08-21). Capillary Ion Chromatography at High Pressure and Temperature. Analytical Chemistry. **84** *(16): 7212–7217.* doi:10.1021/ac301598j.
19. Escuder-Gilabert, L.; Bermúdez-Saldaña, J. M.; Villanueva-Camañas, R. M.; Medina-Hernández, M. J.; Sagrado, S. (2004-04-16). Reliability of the retention factor estimations in liquid chromatography. Journal of Chromatography A. **1033** *(2): 247–255. doi:10.1016/j.chroma.2004.01.038.*
20. Shibukawa, Masami; Shimasaki, Tomomi; Saito, Shingo; Yarita, Takashi (2009-10-01). Superheated Water Ion-Exchange Chromatography: An Experimental Approach for Interpretation of Separation Selectivity in Ion-Exchange Processes. Analytical Chemistry. **81** *(19): 8025–8032.* doi:10.1021/ac9011864.
21. Appling, Dean; Anthony-Cahill, Spencer; Mathews, Christopher (2016). Biochemistry: Concepts and Connections. New Jersey: Pearson. p. 134. ISBN 9780321839923.
22. Alpert, Andrew J.; Hudecz, Otto; Mechtler, Karl (5 May 2015). Anion-Exchange Chromatography of Phosphopeptides: Weak Anion Exchange versus Strong Anion Exchange and Anion-Exchange Chromatography versus Electrostatic Repulsion–Hydrophilic Interaction Chromatography. Analytical Chemistry. **87** *(9): 4704–4711.* doi:10.1021/ac504420c. PMC 4423237. PMID 25827581.
23. Dragan, E. S.; Avram, E.; Dinu, M. V. (1 July 2006). Organic ion exchangers as beads. Synthesis, characterization and applications. Polymers for Advanced Technologies. **17** *(7–8): 571–578. doi:10.1002/pat.755.*
24. Ion Exchange Chromatography Selection Guide. www.pall.com.
25. Ion Exchange Chromatography. www.bio-rad.com.
26. Harris, Daniel C. (2010). Quantitative Chemical Analysis. New York: W. H. Freeman and Company. p. 637. ISBN 1429218150.
27. Harris, Daniel C. (2010). Quantitative Chemical Analysis. New York: W. H. Freeman and Company. p. 638. ISBN 1429218150.
28. Brandt, S (1988). Membrane-based affinity technology for commercial-scale purifications. Bio/Technology. **6** *(7): 779–782.* doi:10.1038/nbt0788-779.

29. Yang, Heewon; Viera, Clarivel; Fischer, Joachim; Etzel, Mark R. (2002). Purification of a Large Protein Using Ion-Exchange Membranes. Industrial & Engineering Chemistry Research. **41** *(6): 1597–1602.* doi:10.1021/ie010585l.

30. Roper, D. Keith; Lightfoot, Edwin N. (1995). Separation of biomolecules using adsorptive membranes. Journal of Chromatography A. **702** *(1–2): 3–26.* doi:10.1016/0021-9673(95)00010-K.

31. Caculitan, Niña G.; Kai, Hiroyuki; Liu, Eulanca Y.; Fay, Nicole; Yu, Yan; Lohmüller, Theobald; O'Donoghue, Geoff P.; Groves, Jay T. (2014). Size-Based Chromatography of Signaling Clusters in a Living Cell Membrane. Nano Letters. **14** *(5): 2293–2298.* doi:10.1021/nl404514e.

32. Svec, Frantisek.; Frechet, Jean M. J. (1992). Continuous rods of macroporous polymer as high-performance liquid chromatography separation media. Analytical Chemistry. **64** *(7): 820–822.* doi:10.1021/ac00031a022.

33. Garrett, Reginald H.; Grisham, Charles M. (2009). Biochemistry (4th ed.). Pacific Grove, Calif.: Brooks/Cole. pp. 71–75. ISBN 978-0-495-11464-2.

34. Dasgupta, Purnendu K. (1992). Ion Chromatography the State of the Art. Analytical Chemistry. **64** *(15): 775A.* doi:10.1021/ac00039a722.

35. Sakash, J.B.; Kantrowitz, E.R. (2000). The contribution of individual interchain interactions to the stabilization of the T and R states of Escherichia coli aspartate transcarbamoylase.. *J* Biol Chem. **275** *(37): 28701–7. doi:10.1074/jbc.M005079200. PMID 10875936.*

36. Fairhead, M. (2013). Plug-and-Play Pairing via Defined Divalent Streptavidins. J Mol Biol. **426** *(1): 199–214. doi:10.1016/j.jmb.2013.09.016. PMC 4047826. PMID 24056174.*

37. Heftmann, Erich (2004). Chromatography. Fundamentals and Applications of Chromatography and Related Differential Migration Methods: Applications. Amsterdam: Elsevier. p. 716. ISBN 9780080472256.

38. Gregor, Harry P. (1951). Gibbs-Donnan Equilibria in Ion Exchange Resin Systems. Journal of the American Chemical Society. **73** *(2): 642–650.* doi:10.1021/ja01146a042.

39. Rieman, William, Harold F. Walton, R. Belcher, and H. Freiser (2013). Ion Exchange in Analytical Chemistry International Series of Monographs in Analytical Chemistry. Burlington: Elsevier Science.

40. Hauser, Peter C. (2016). Chapter 2. Determination of Alkali Ions in Biological and Environmental Samples. In Astrid, Sigel; Helmut, Sigel; Roland K.O., Sigel. The Alkali Metal Ions: Their Role in Life. Metal Ions in Life Sciences. **16**. Springer. pp. 11–25. doi:10.1007/978-3-319-21756-7_2.

41. Luqman, Mohammad, and Inamuddin (2012). Ion Exchange Technology II. Springer Netherlands. p. 169. ISBN 978-94-007-4026-6.

42. Robert E. Smith (31 December 1987). Ion Chromatography Applications. CRC Press. ISBN 978-0-8493-4967-6.

43. Bhattacharyya, Lokesh; Rohrer, Jeffrey (2012). Applications of Ion

Chromatography in the Analysis of Pharmaceutical and Biological Products. Wiley. p. 247. ISBN 0470467096.

44. Hanko, Valoran P.; Rohrer, Jeffrey S. (2012). Ion Chromatography Analysis of Aminoglycoside Antibiotics. Applications of Ion Chromatography for Pharmaceutical and Biological Products. p. 175. doi:10.1002/9781118147009.ch8. ISBN 9781118147009.

45. Jenke, D. (2011). Application of Ion Chromatography in Pharmaceutical and Drug Analysis. Journal of Chromatographic Science. **49** *(7): 524–39. doi:10.1093/* chrsci/49.7.524. PMID 21801484.

46. Garrett, Reginald H.; Grisham, Charles M. (2013). Biochemistry (5th ed.). Belmont, CA: Brooks/Cole, Cengage Learning. p. 108. ISBN 9781133106296.

47. Paul-Dauphin, Stephanie; Morgan, Millan-Agorio; Herod, Kandiyoti (6 October 2007). Probing Size Exclusion Mechanisms of Complex Hydrocarbon Mixtures: The Effect of Altering Eluent Compositions. Energy & Fuels. 6. **21** *(6): 3484–3489.* doi:10.1021/ef700410e. Retrieved October 6, 2007.

48. Skoog, D. A.; Principles of Instrumental Analysis, 6th ed.; Thompson Brooks/ Cole: Belmont, CA, 2006, Chapter 28.

49. Rouessac, Annick; Rouessac, Francis (2000). Chemical analysis : modern instrumental methods and techniques (Engl. ed.). Chichester [u.a.]: Wiley. pp. 101–103. ISBN 0471972614.

50. Ballou, David P.; Benore, Marilee; Ninfa, Alexander J. (2008). Fundamental laboratory approaches for biochemistry and biotechnology. (2nd ed.). Hoboken, N.J.: Wiley. p. 129. ISBN 9780470087664.

51. Lathe, GH and Ruthven, CR (1955) The separation of substances on the basis of their molecular weights, using columns of starch and water. Biochem J. 60(4):xxxiv.

52. Lathe, GH and Ruthven, CR (1956) The separation of substances and estimation of their relative molecular sizes by the use of columns of starch in water. Biochem. J. 62(4): 665-674. article

53. John Scott Award

54. Porath, J and Flodin, P (1959) Gel filtration: A method for desalting and group separation. Nature 183(4676): 1657-1659.

55. Eisenstein, M (2006) A look back, adventures in the matrix. Nature Methods 3(5): 410 article

56. Moore, J. C. (1964) Gel Permeation Chromatography. 1. A New Method for Molecular Weight Distribution of High Polymers. Journal of Polymer Science: Part A 2: 835-843

57. Striegel, A. M.; Kirkland, J. J.; Yau, W. W.; Bly, D. D.; Modern Size Exclusion Chromatography, Practice of Gel Permeation and Gel Filtration Chromatography, 2nd ed.; Wiley: NY, 2009.

58. Grubisic, Z.; Rempp, P.; Benoit, H. (1967) A universal calibration for Gel Permeation Chromatography. Journal of Polymer Science: Polym. Lett. 5: 753-759

59. Sun, T.; Chance, R. R.; Graessley, W. W.; and Lohse, D. J. (2004) A study of the separation principle in Size Exclusion Chromatography. Macromolecules 37: 4304-4312.

60. Wang, Y.; Teraoka, I.; Hansen, F. Y.; Peters, G. H.; and Hassager, O. (2010) A theoretical study of the separation principle in Size Exclusion Chromatography. Macromolecule 43: 1651-1659.

61. Polymer Science Learning Center (PSLC) - Size Exclusion Chromatography

62. Ballou, David P.; Benore, Marilee; Ninfa, Alexander J. (2008). Fundamental laboratory approaches for biochemistry and biotechnology. (2nd ed.). Hoboken, N.J.: Wiley. p. 127. ISBN 9780470087664.

CHAPTER 7

Special Techniques of Chromatography

Chapter Contents

- **Reversed-Phase Chromatography**
- **Stationary Phases**
- **Mobile Phases**
- **Simulated Moving Bed**
- **Countercurrent Chromatography**

Chromatography can be carried out in many ways. In column chromatography and in thin layer chromatography (TLC) a solution of the mixture flows over a solid adsorbent. Separation occurs as molecules are adsorbed and desorbed while passing over the surface. In paper chromatography applications, the mixture is partitioned between water molecules adsorbed on the paper and a solvent that moves over the paper. In gas chromatography (GC, also called vapour phase chromatography), a mixture of volatile compounds is separated by passing the vapour over an adsorbent packing in a long, heated tube

REVERSED-PHASE CHROMATOGRAPHY

Reversed-phase chromatography (also called RPC, reverse-phase chromatography, or hydrophobic chromatography) includes any chromatographic method that uses a hydrophobic stationary phase.RPC refers to liquid (rather than gas) chromatography.

The term reversed-phase has a historical background. In the 1970s, most liquid chromatography was performed using a solid support stationary phase (also called a "column") containing unmodified silica or alumina resins. This method is now called "normal phase chromatography". In normal phase chromatography, the stationary phase is hydrophilic and therefore has a strong affinity for hydrophilic molecules in the mobile phase. Thus, the hydrophilic molecules in the mobile phase tend to bind (or adsorb) to the column, while the hydrophobic molecules pass through the column and are eluted first. In normal phase chromatography, hydrophilic molecules can be eluted from the column by increasing the polarity of the solution in the mobile phase.

The introduction of a technique using alkyl chains covalently bonded to the solid support created a hydrophobic stationary phase, which has a stronger affinity for hydrophobic compounds. The use of a hydro*phobic* stationary phase can be considered the opposite, or "reverse", of normal phase chromatography - hence the term "reversed-phase chromatography". Reversed-phase chromatography employs a polar (aqueous) mobile phase. As a result, hydrophobic molecules in the polar mobile phase tend to adsorb to the hydrophobic stationary phase, and hydro*philic* molecules in the mobile phase will pass through the column and are eluted first. Hydrophobic molecules can be eluted from the column by decreasing the polarity of the mobile phase using an organic (non-polar) solvent, which reduces hydrophobic interactions. The more hydrophobic the molecule, the more strongly it will bind to the stationary phase, and the higher the concentration of organic solvent that will be required to elute the molecule.

Many of the mathematical and experimental considerations used in other chromatographic methods also apply to RPC (for example, the separation resolution is dependent on the length of the column). It can be used for the separation of a wide variety of molecules. It is not typically used for separation of proteins, because the organic solvents used in RPC can denature many proteins. For this reason, normal phase chromatography is more commonly used for separation of proteins.

Today, RPC is a frequently used analytical technique. There are a variety of stationary phases available for use in RPC, allowing great flexibility in the development of separation methods.

STATIONARY PHASES

Silica-based Stationary Phases

Any inert non-polar substance that achieves sufficient packing can be used for reversed-phase chromatography. The most popular column is an octadecyl carbon chain (C18)-bonded silica (USP classification L1) with 297 columns commercially available. This is followed by C8-bonded silica (L7 - 166 columns), pure silica (L3 - 88 columns), cyano-bonded silica (L10 - 73 columns) and phenyl-bonded silica (L11 - 72 columns). Note that C18, C8 and phenyl are dedicated reversed-phase resins, while cyano columns can be used in a reversed-phase mode depending on analyte and mobile phase conditions. It should be noted at this point that not all C18 columns have identical retention properties. Surface functionalization of silica can be performed in a monomeric or a polymeric reaction with different short-chain organosilanes used in a second step to cover remaining silanol groups (end-capping). While the overall retention mechanism remains the same, subtle differences in the surface chemistries of different stationary phases will lead to changes in selectivity.

Modern columns have different polarity. PFP is pentafluorphenyl. CN is cyano. NH2 is amino. ODS is octadecyl or C18. ODCN is a mixed mode column consisting of C18 and nitrile. SCX is strong cationic exchange (used for separation of organic amines). SAX is strong anionic exchange (used for separation of carboxylic acid compounds).

MOBILE PHASES

Mixtures of water or aqueous buffers and organic solvents are used to elute analytes from a reversed-phase column. The solvents must be miscible with water, and the most common organic solvents used are acetonitrile, methanol, and tetrahydrofuran (THF). Other solvents can be used such as ethanol or 2-propanol (isopropyl alcohol). Elution can be performed isocratically (the water-solvent composition does not change during the separation process) or by using a solution gradient (the water-solvent composition changes during the separation process, usually by decreasing the polarity). The pH of the mobile phase can have an important role on the retention of an analyte and can change the selectivity of certain analytes.

Charged analytes can be separated on a reversed-phase column by the use of ion-pairing (also called ion-interaction). This technique is known as reversed-phase ion-pairing chromatography.

SIMULATED MOVING BED

In manufacturing, the **simulated moving bed** (SMB) process is a highly engineered process for implementing chromatographic separation. It is used to separate one chemical compound or one class of chemical compounds from one or more other chemical compounds to provide significant quantities of the purified or enriched material at a lower cost than could be obtained using simple (batch) chromatography. It cannot provide any separation or purification that cannot be done by a simple column purification. The process is rather complicated. The single advantage which it brings to a chromatographic purification is that it allows the production of large quantities of highly purified material at a dramatically reduced cost. The cost reductions come about as a result of: the use of a smaller amount of chromatographic separation media stationary phase, a continuous and high rate of production, and decreased solvent and energy requirements. This improved economic performance is brought about by a valve-and-column arrangement that is used to lengthen the stationary phase indefinitely and allow very high solute loadings to the process.

In the conventional **moving bed technique** of production chromatography the feed entry and the analyte recovery are simultaneous and continuous, but because of practical difficulties with a continuously moving bed, simulated moving bed technique was proposed. In the **simulated moving bed technique** instead of moving the bed, the feed inlet, the solvent or eluent inlet and the desired product exit and undesired product exit positions are moved continuously, giving the impression of a moving bed, with continuous flow of solid particles and continuous flow of liquid in the opposite direction of the solid particles.

True moving bed chromatography (TMBC) is only a theoretical concept. Its simulation, **SMBC** is achieved by the use of a multiplicity of columns in series and a complex valve arrangement, which provides for flow of the feed mixture and solvent, and "eluent or desorbent feed at any column. The valving and piping arrangements and the predetermined control of these allow switching at regular intervals the sample entry in one direction, the solvent entry in the same direction but at a different location in the continuous loop, whilst changing the *fast product* and *slow product* takeoff positions to also move in the same direction, but at different relative locations within the loop.

Ref 3 explains that the advantage of the SMBC is high production rate, because a system could be near continuous, whilst its disadvantage is that it only performs one cut in mixtures. Thus, it is well-suited for separation of a binary mixture. With multiple cuts, analogous to a series of distillation columns, multiple compounds can be separated from a mixture of more than two compounds. With regard to efficiency it compares with simple chromatography technique like continuous distillation does with batch distillation.

Construction

Specifically, an SMB system has two or more identical columns, which are connected to the mobile phase pump, and each other, by a multi-port valve. The plumbing is configured in such a way that:

- all columns will be connected in series, forming a single continuous loop;
- typically, between each column there will be provisions for four process streams: incoming feed mixture, exiting purified fast component, exiting purified slow component, and incoming solvent or eluent;

and

- each process stream (two inlets and two outlets) will proceed in the same direction after a set time (the steptime).

Advantages

SMB provides lower production cost by requiring less column volume, less chromatographic separation media ("packing" or "stationary phase"), using less solvent and less energy, and requiring far less labor.

At industrial scale an SMB chromatographic separator is operated continuously, requiring less resin and less solvent than batch chromatography. The continuous operation facilitates operation control and integration into production plants. Low eluent consumption High product concentration High productivity Continuous process This system is useful in the supercritical fluid extraction to obtain large quantity of specific product.

Drawbacks

The drawbacks of the SMB are higher investment cost compared to single column operations, a higher complexity, as well as higher maintenance costs. But these drawbacks are effectively compensated by the better yield and a much lower solvent consumption as well as a much higher productivity compared to simple batch separations.

For purifications, in particular the isolation of an intermediate single component or a fraction out of a multicomponent mixture, the SMB is not as ideally suited. Normally, a single SMB will separate only two fractions from each other, but a series or "train" of SMBs can perform multiple cuts and purify one or more products from a multi-component mixture. SMB is not readily suited for solvent gradients. Solvent gradient purification may be preferred for the purification of some biomolecules. A continuous chromatography technique to overcome the two fraction limit and to apply gradients is multicolumn countercurrent solvent gradient purification (MCSGP).

Applications

In size exclusion chromatography, where the separation process is driven by entropy, it is not possible to increase the resolution attained by a column via temperature or solvent gradients. Consequently, these separations often require SMB, to create usable retention time differences between the molecules or particles being resolved. SMB is also very useful in the pharmaceutical industry, where resolution of molecules having different chirality must be done on a very large scale. For the purification of fructose, e.g. in high fructose corn syrup, or amino-acids, biological-acids, etc. on an industrial scale simulated moving bed chromatography is used.

COUNTERCURRENT CHROMATOGRAPHY

Countercurrent chromatography (**CCC**, also **counter-current chromatography**) is a form of liquid–liquid chromatography that uses a liquid stationary phase that is held in place by centrifugal force and is used to separate, identify, and quantify the chemical components of a mixture. In its broadest sense, countercurrent chromatography encompasses a collection of related liquid chromatography techniques that employ two immiscible liquid phases without a solid support. The two liquid phases come in contact with each other as at least one phase is pumped through a column, a hollow tube or a series of chambers connected with channels, which contains both phases. The resulting dynamic mixing and settling action allows the components to be separated by their respective solubilities in the two phases. A wide variety of two phase solvent systems consisting of at least two immiscible liquids may be employed to provide the proper selectivity for the desired separation.

Some types of countercurrent chromatography, such as dual flow CCC, feature a true countercurrent process where the two immiscible phases flow past each other and exit at opposite ends of the column. More often, however, one liquid acts as the stationary phase and is retained in the column while the mobile phase is pumped through it. The liquid stationary phase is held in place by gravity or by centrifugal force. An example of a gravity method is called droplet counter current chromatography (DCCC). There are two modes by which the stationary phase is retained by centrifugal force: hydrostatic and hydrodynamic. In the hydrostatic method, the column, a series of chambers connected by channels, is rotated about a central axis. Hydrostatic instruments are marketed under the name centrifugal partition chromatography (CPC). Hydrodynamic instruments are often marketed as high-speed or high-performance countercurrent chromatography (HSCCC and HPCCC respectively) instruments which rely on the Archimedes` screw force in a helical coil to retain the stationary phase in the column.

The components of a CCC system are similar to most liquid chromatography configurations such as high-performance liquid chromatography. One or more pumps deliver the phases to the column which is the CCC instrument itself. Samples are introduced into the column through a sample loop filled with an automated or manual syringe. The outflow is monitored with various detectors such as ultraviolet–visible spectroscopy or mass spectrometry. The operation of the pumps, CCC instrument, sample injection, and detection may be controlled manually or with a microprocessor.

History

The predecessor of modern countercurrent chromatography theory and practice was countercurrent distribution (CCD). The theory of CCD was described in the 1930s by Randall and Longtin. Archer Martin and Richard Laurence Millington Synge developed the methodology further during the 1940s.Finally, Lyman C. Craig introduced the Craig countercurrent distribution apparatus in 1944 which made CCD practical for laboratory work. CCD was used to separate a wide variety of useful compounds for several decades.

Support-free Liquid Chromatography

Standard column chromatography consists of a solid stationary phase and a liquid mobile phase, while gas chromatography (GC) uses a solid or liquid stationary phase on a solid support and a gaseous mobile phase. By contrast, in liquid-liquid chromatography, both the mobile and stationary phases are liquid. The contrast is, however not as stark as it first appears. In reversed-phase chromatography, for example, the stationary phase can be regarded as a liquid which is immobilized by chemical bonding to a micro-porous silica solid support. In countercurrent chromatography centrifugal or gravitational forces immobilize the stationary liquid layer. By eliminating solid supports, permanent adsorption of the analyte onto the column is avoided, and a high recovery of the analyte can be achieved. The countercurrent chromatography instrument is easily switched between normal phase chromatography and reversed-phase chromatography simply by changing the mobile and stationary phases. With column chromatography, the separation potential is limited by the commercially available stationary phase media and its particular characteristics. Nearly any pair of immiscible solutions can be used in countercurrent chromatography provided that the stationary phase can be successfully retained.

Solvent costs are also generally lower than for high-performance liquid chromatography (HPLC). In comparison to column chromatography, flows and total solvent usage can in most countercurrent chromatography separations may be reduced by half and even up to a tenth. Also, the cost of purchasing and disposing of stationary phase media is eliminated. Another advantage of countercurrent chromatography is that experiments conducted in the laboratory can be scaled to industrial volumes. When gas chromatography or HPLC is carried out with large volumes, resolution is lost due to issues with surface-to-volume ratios and flow dynamics; this is avoided when both phases are liquid.

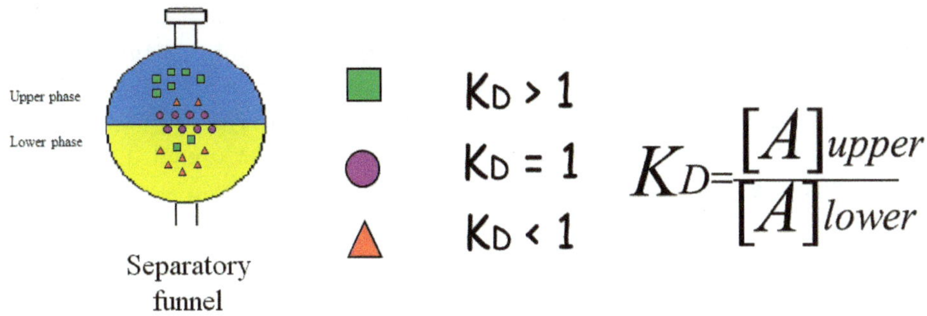

Partition coefficient (K_D).

The CCC separation process can be thought of as occurring in three stages: mixing, settling, and separation of the two phases (although they often occur continuously). Vigorous mixing of the phases is critical in order to maximize the interfacial area between them and enhance mass transfer. The analyte will distribute between the phases according to its partition coefficient which is also called the distribution coefficient, distribution constant, or partition ratio and is represented by P, K, D, K_c, or K_D. The partition coefficient for an analyte in a particular biphasic solvent system is independent of the volume of the instrument, flow rate, stationary phase retention volume ratio and the g-force required to immobilize the stationary phase. The degree of stationary phase retention is a crucial parameter. Common factors that influence stationary phase retention are flow rate, solvent composition of the biphasic solvent system, and the g-force. The stationary phase retention is represented by the stationary phase volume retention ratio (Sf) which is the volume of the stationary phase divided by the total volume of the instrument. The settling time is a property of the solvent system and the sample matrix, both of which greatly influence stationary phase retention.

To most process chemists, the term countercurrent implies two immiscible liquids moving in opposing directions, as typically occurs in large centrifugal extractor units. With the exception of dual flow (see below) CCC, most countercurrent chromatography modes of operation have a stationary phase and a mobile phase. Even in this situation, countercurrent flows occur within the instrument column. Several researchers have proposed renaming both CCC & CPC to liquid-liquid chromatography, but others feel the term countercurrent itself is a misnomer.

Unlike column chromatography and high-performance liquid chromatography, countercurrent chromatography operators can inject large volumes relative to column volume. Typically 5 to 10% of coil volume can be injected. In some cases this can be increased to as high as 15 to 20% of the coil volume. Typically, most modern commercial CCC and CPC can inject 5 to 40 g per liter capacity. The range is so large, even for a specific instrument, let alone all instrument options, as the type of target, matrix and available biphasic solvent vary so much. Approximately 10 g per liter would be a more typical value, that the majority of applications could use as a base value.

The countercurrent separation starts with choosing an appropriate biphasic solvent system for the desired separation. A wide array of biphasic solvent mixtures are available to the CCC practitioner including the combination n-hexane (or heptane), ethyl acetate, methanol and water in different proportions. This basic solvent system is sometimes referred to as the HEMWat solvent system. The choice of solvent system may be guided by perusal of the CCC literature. The familiar technique of thin layer chromatography may also be employed to determine an optimal solvent system. The organization of solvent systems into "families" has greatly facilitated the choice of solvent systems as well. A solvent system can be tested with a one-flask partitioning experiment. The measured partition coefficient from the partitioning experiment will indicate the elution behavior of the compound. Typically, it is desirable to choose a solvent system where the target compound(s) have a partition coefficient between 0.25 and 8. Historically, it was thought that no commercial countercurrent chromatograph could cope with the high viscosities of ionic liquids. However, modern instruments that can accommodate 30 to 70+ % ionic liquids (and potentially 100% ionic liquid, if both phases are suitably customized ionic liquids) have become available. Ionic liquids can be customized for polar / non-polar organic, achiral and chiral compounds, bio-molecule, and inorganic separations, as ionic liquids can be customized to have extraordinary solvency and specificity.

After the biphasic solvent system has been chosen a batch of is formulated and equilibrated in a separatory funnel. This step is called pre-equilibration of the solvent system. The two phases are separated. Then the column is filled with stationary with a pump. Next, the column is set an equilibration conditions, such as the desired rotation speed, and the mobile phase is pumped through the column. The mobile phase displaces the a portion of the stationary phase until column equilibration is achieved and the mobile phase elutes from the column. The sample may be introduced into the column at any time during the column equilibration step or after equilibration has been accomplished. After the volume of eluant surpasses the volume of the mobile phase in the column, the sample components will begin to elute. Compounds with a partition coefficient of unity will elute when one column volume of mobile phase has passed through the column since the time of injection. The flow is stopped after the target compound(s) are eluted or the column is extruded by pumping the stationary phase through the column. An example of a major application of countercurrent chromatography is to take an extremely complex matrix such as a plant extract, perform the countercurrent chromatography separation with a carefully selected solvent system, and extrude the column to recover all of the sample. The original complex matrix will have been fractionated into discrete narrow polarity bands, which may then be assayed for chemical composition or bioactivity. Performing one or more countercurrent chromatography separations in conjunction with other chromatographic and non chromatographic techniques has the potential for rapid advances in compositional recognition of extremely complex matrices.

Droplet CCC

Droplet countercurrent chromatography (DCCC) was introduced in 1970 by Tanimura, Pisano, Ito, and Bowman. DCCC uses only gravity to move the mobile phase through the stationary phase which is held in long vertical tubes connected in series. In the descending mode, droplets of the denser mobile phase and sample are allowed to fall through the columns of the lighter stationary phase using only gravity. If a less dense mobile phase is used it will rise through the stationary phase; this is called ascending mode. The eluent from one column is transferred to another; the more columns that are used, the more theoretical plates can be achieved. DCCC enjoyed some success with natural product separations but was largely eclipsed by the rapid development of high-speed countercurrent chromatography. The main limitation of DCCC is that flow rates are low, and poor mixing is achieved for most binary solvent systems.

Hydrodynamic CCC

The modern era of CCC began with the development by Dr. Yoichiro Ito of the planetary centrifuge which was first introduced in 1966 as a closed helical tube which was rotated on a "planetary" axis as is turned on a "sun" axis. A flow-through model was subsequently developed and the new technique was called countercurrent chromatography in 1970. The technique was further developed by employing test mixtures of DNP amino acids in chloroform:glacial acetic acid:0.1 M aqueous hydrochloric acid (2:2:1 v/v) cosolvent system. Much development was needed to engineer the instrument so that required planetary motion could be sustained while the phases were being pumped through the coil(s).Parameters such as the relative rotation of the two axes (synchronous or non-synchronous), the direction of flow through the coil, and the rotor angles were investigated.

High-speed

By 1982 the technology was sufficiently advanced for the technique to be called "high-speed" countercurrent chromatography (HSCCC). Peter Carmeci initially commercialized the PC Inc. Ito Multilayer Coil Separator/Extractor which utilized a single bobbin (onto which the coil is wound) and a counterbalance, plus a set of "flying leads" which are tubing that connect the bobbins. Dr. Walter Conway & others later evolved the bobbin design such that multiple coils, even coils of different tubing sizes, could be placed on the single bobbin. Edward Chou later evolved and commercialized a triple bobbin design as the Pharmatech CCC which had a de-twist mechanism for leads between the three bobbins. The Quattro CCC released in 1993 further evolved the commercially available instruments by utilizing a novel mirror image, twin bobbin design that did not need the de-twist mechanism of the Pharmatech between the multiple bobbins, so could still accommodate multiple bobbins on the same instrument. Hydrodynamic CCC are now available with up to 4 coils per instrument. These coils can be in PTFE, PEEK, PVDF, or stainless steel tubing. The 2, 3 or 4 coils can all be of the same bore to facilitate "2D" CCC (see below). The coils may be connected in series to lengthen the coil and increase the capacity, or the coils may be linked in parallel so that 2, 3, or 4 separations may be done simultaneously. The coils can also be of different sizes, on one instrument, ranging from 1 to 2 to 3.7 to 6 mm on one instrument, thus allowing a single instrument to optimize from mg to kilos per day. More recently instrument derivatives have started to be offered with rotating seals for various hydrodynamic CCC designs, instead of flying leads, either as custom options or as standard.

High-Performance

The operating principle of CCC equipment requires a column consisting of a tube coiled around a bobbin. The bobbin is rotated in a double-axis gyratory motion (a cardioid), which causes a variable g-force to act on the column during each rotation. This motion causes the column to see one partitioning step per revolution and components of the sample separate in the column due to their partitioning coefficient between the two immiscible liquid phases. "High-performance" countercurrent chromatography (HPCCC) works in much the same way as HSCCC. A seven-year R&D process produced HPCCC instruments that generated 240 g's, compared to the 80 g's of the HSCCC machines. This increase in g-force and larger bore of the column has enabled a tenfold increase in throughput, due to improved mobile phase flow rates and a higher stationary phase retention. Countercurrent chromatography is a preparative liquid chromatography technique, however with the advent of the higher-g HPCCC instruments it is now possible to operate instruments with sample loadings as low as a few milligrams, whereas in the past 100s of milligrams had been necessary. Major application areas for this technique include natural products purification and also drug development.

Hydrostatic

Hydrostatic CCC or centrifugal partition chromatography (CPC) was invented in the 1980s by the Japanese company Sanki Engineering Ltd, whose president was Kanichi Nunogaki. CPC has been extensively developed in France starting from the late 1990s. In France, they initially optimized the stacked disc concept initiated by Sanki. More recently, in France and UK, non-stacked disc CPC configurations have been developed with PTFE, stainless steel or titanium rotors. These have been designed to overcome possible leakages between the stacked discs of the original concept, and to allow steam cleaning for Good manufacturing practice. The volumes ranging from a 100 ml to 12 liters are available in different rotor materials. The 25 liter rotor CPC has a titanium rotor. This technique is sometimes sold under the name fast CPC or high-performance CPC.

Principle

CPC is based on the principles of countercurrent chromatography: the stationary phase of a two-phase system is maintained in the instrument under a centrifugal force while the mobile phase is pumped through a series of chambers or cells. The stationary phase is retained inside the rotor by the centrifugal force generated by rotation around a single axis. Discrete chambers connected by channels are the sites of mixing and settling action that allows the compounds being separated to distribute between the two phases. As the mobile phase exits the columns it is collected in test tubes for further analysis. Chemical compounds subjected to CPC are separated based on the partition coefficient of each compound in this two-phase system.

Realization

The centrifugal partition chromatograph instrument is constituted with a unique rotor which contains the column. This rotor rotates on its central axis (while HSCCC column rotates on its planetary axis and simultaneously rotates eccentrically about another solar axis). With less vibrations and noise, the CPC offers a typical rotation speed range from 500 to 2000 rpm. Contrary to hydrodynamic CCC, the rotation speed is not directly proportional to the retention volume ratio of the stationary phase. Like DCCC, CPC can be operated in either descending or ascending mode, where the direction is relative to the force generated by the rotor rather than gravity. A redesigned CPC column with larger chambers and channels has been named centrifugal partition extraction (CPE). In the CPE design, faster flow rates and increased column loading can be achieved.

Advantages

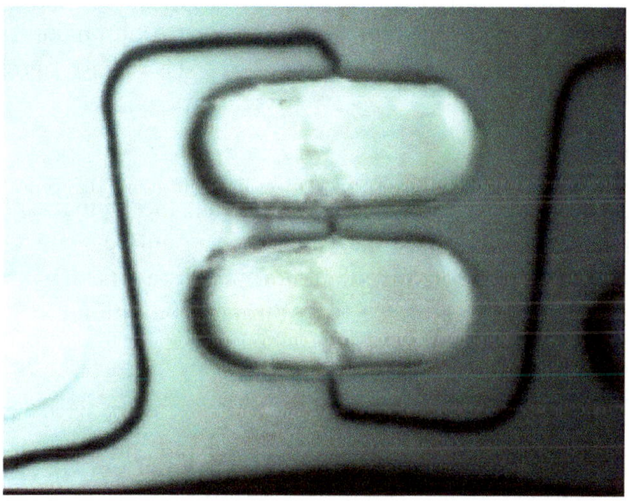

Visualization of mixing and settling in a CPC twin cell.

The CPC offers now the direct scale-up from the analytical apparatuses (few milliliters) to industrial apparatuses (some liters) for fast batch production. The CPC seems particularly suited to accommodate aqueous two-phase solvent systems.Generally, CPC instruments can retain solvent systems that are not well-retained in a hydrodynamic instrument due to small differences in density between the phases. It has been very helpful for the development of CPC instrumentation to visualize the flow patterns which give rise to the mixing and settling in the CPC chamber with an asynchronous camera and a stroboscope triggered by the CPC rotor.

Modes of Operation

The aforementioned hydrodynamic and hydrostatic instruments may be employed in a variety of ways, or modes of operation, in order to address the particular separation needs of the scientist. Many modes of operation have been devised to take advantage of the strengths and potentialities of the countercurrent chromatography technique. Generally, the following modes may be performed with commercially available instruments.

Normal-phase

Normal phase elution is achieved by pumping the non-aqueous or phase of a biphasic solvent system through the column as the mobile phase, with a more polar stationary phase being retained in the column. The cause of original nomenclature of is relevant. As original stationary phases of paper chromatography were superseded by more efficient materials such as diatomaceous earths (natural micro-porous silica) and followed by modern silica gel, the thin-layer chromatography stationary phase was polar (hydroxy groups attached to silica) and maximum retention was achieved with non-polar solvents such as n-hexane. Progressively more polar eluents were then used to move polar compounds up the plate. Various alkane bonded phases were tried with C18 becoming the most popular. Alkane chains were chemically bonded to the silica, and a reversal of the elution trend occurred. Thus a polar stationary became "normal" phase chromatography, and the non-polar stationary phase chromatography became "reversed" phase chromatography.

Reversed-phase

In reversed-phase (e.g. aqueous mobile phase) elution, the aqueous phase is used as the mobile phase with a less polar stationary phase. In countercurrent chromatography the same solvent system may be used in either normal or reversed phase mode simply by switching the direction of mobile phase flow through the column.

Elution-extrusion

The extrusion of stationary phase from the column at the end of a separation experiment by stopping rotation and pumping solvent or gas through the column was used by CCC practitioners before the term EECCC was suggested. In elution-extrusion mode (EECCC), The mobile phase is extruded after a certain point by switching the phase being pumped into the system whilst maintaining rotation. For example, if the separation has been initiated with the aqueous phase as the mobile phase at a certain point the organic phase is pumped through the column which effectively pushes out both phases that are present in the column at the time of switching. The complete sample is eluted in the order of polarity (either normal or reversed) without loss of resolution by diffusion. It requires only one column volume of solvent phase and leaves the column full of fresh stationary phase for the subsequent separation.

Gradient

The use of a solvent gradient is very well developed in column chromatography but is less common in CCC. A solvent gradient is produced by increasing (or decreasing) the polarity of the mobile phase during the separation to achieve optimal resolution across a wider range of polarities. For example, a methanol-water mobile phase gradient may be employed using heptane as the stationary phase. This is not possible with all biphasic solvent systems, due to excessive loss of stationary phase created by disruption the equilibrium conditions within the column. Gradients may either be produced in steps, or continuously.

Dual-mode

In dual-mode, the mobile and stationary phases are reversed part way through the separation experiment. This requires changing the phase being pumped through the column as well as the direction of flow. Dual-mode operation is likely to elute the entire sample from the column but the order of elution is disrupted by switching the phase and direction of flow.

Dual-flow

Dual-flow, also known as dual, countercurrent chromatography occurs when both phases are flowing in opposite directions inside the column. Instruments are available for dual-flow operation for both Hydrodynamic and hydrostatic CCC. Dual-flow countercurrent chromatography was first described by Yoichiro Ito in 1985 for foam CCC where gas-liquid separations were performed. Liquid–liquid separations soon followed. The countercurrent chromatography instrument must be modified so that both ends of the column have both inlet and outlet capabilities. This mode may accommodate continuous or sequential separations with the sample being introduced in the middle of the column or between two bobbins in a hydrodynamic instrument. A technique called intermittent countercurrent extraction (ICcE) is a quasi-continuous method where the flow of the phases is alternated “intermittently” between normal and reversed-phase elution so that the stationary phase also alternates.

Recycling and Sequential

Recycling chromatography is mode practiced in both HPLC and CCC. In recycling chromatography, the target compounds are reintroduced into the column after they elute. Each pass through the column increases the number of theoretical plates the compounds experience and enhances chromatographic resolution. Direct recycling must be done with an isocratic solvent system. With this mode, the eluant can be selectively re-chromatographed on the same or a different column in order to facilitate the separation. This process of selective recycling has been termed a heart-cut and is especially effective in purifying selected target compounds with some sacrificial loss of recovery. The process of re-separating selected fractions from one chromatography experiment with another chromatographic method has long been practiced by scientists. Recycling and sequential chromatography is a streamlined version of this process. In CCC, the separation characteristics of the column may be modified simply by changing the composition of the biphasic solvent system.

Ion-exchange and pH-Zone-refining

In an conventional CCC experiment the biphasic solvent system is pre-equilibrated before the instrument is filled with the stationary phase and equilibrated with the mobile phase. An ion-exchange mode has been created by modifying both of the phases after pre-equilibration. Generally, an ionic displacer (or eluter) is added to mobile phase and an ionic retainer is added to the stationary phase. For example, the aqueous mobile phase may contain NaI as a displacer and the organic stationary phase may be modified with the quaternary ammonium salt called Aliquat 336 as a retainer. The mode known a pH-zone-refining is a type of ion-exchange mode that utilizes acids and/or bases as solvent modifiers.Typically, the analytes are eluted in an order determined by their pKa values. For example, 6 oxindole alkaloids were isolated from a 4.5g sample of *Gelsemium elegans* stem extract with a biphasic solvent system composed of hexane–ethyl acetate–methanol–water (3:7:1:9, v/v) where 10 mM triethylamine (TEA) was added to the upper organic stationary phase as a retainer and 10 mM hydrochloric acid (HCl) to the aqueous mobile phase as an eluter. Ion-exchange modes such as pH-zone-refining have tremendous potential because high sample loads can be achieved without sacrificing separation power. It works best with ionizable compounds such as nitrogen containing alkaloids or carboxylic acid containing fatty acids.

Hydrophobic Interaction Chromatography

Hydrophobic interactions between proteins and the chromatographic matrix can be exploited to purify proteins. In hydrophobic interaction chromatography the matrix material is lightly substituted with hydrophobic groups. These groups can range from methyl, ethyl, propyl, octyl, or phenyl groups.[] At high salt concentrations, non-polar sidechains on the surface on proteins "interact" with the hydrophobic groups; that is, both types of groups are excluded by the polar solvent (hydrophobic effects are augmented by increased ionic strength). Thus, the sample is applied to the column in a buffer which is highly polar. The eluant is typically an aqueous buffer with decreasing salt concentrations, increasing concentrations of detergent (which disrupts hydrophobic interactions), or changes in pH.

In general, Hydrophobic Interaction Chromatography (HIC) is advantageous if the sample is sensitive to pH change or harsh solvents typically used in other types of chromatography but not high salt concentrations. Commonly, it is the amount of salt in the buffer which is varied. In 2012, Müller and Franzreb described the effects of temperature on HIC using Bovine Serum Albumin (BSA) with four different types of hydrophobic resin. The study altered temperature as to effect the binding affinity of BSA onto the matrix. It was concluded that cycling temperature from 50 degrees to 10 degrees would not be adequate to effectively wash all BSA from the matrix but could be very effective if the column would only be used a few times. Using temperature to effect change allows labs to cut costs on buying salt and saves money.

If high salt concentrations along with temperature fluctuations want to be avoided you can use a more hydrophobic to compete with your sample to elute it. [source] This so-called salt independent method of HIC showed a direct isolation of Human Immunoglobulin G (IgG) from serum with satisfactory yield and used Beta-cyclodextrin as a competitor to displace IgG from the matrix. This largely opens up the possibility of using HIC with samples which are salt sensitive as we know high salt concentrations precipitate proteins.

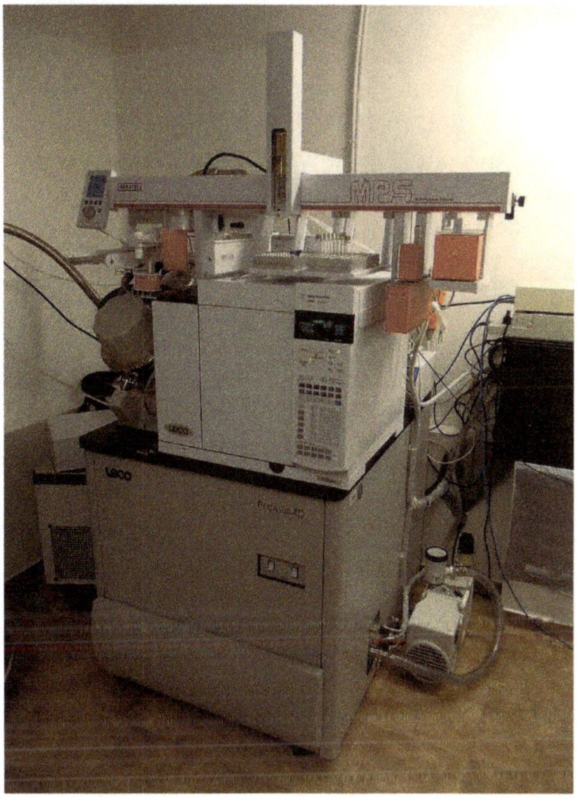

Two-dimensional chromatograph GCxGC-TOFMS at Chemical Faculty of GUT Gdańsk, Poland, 2016.

Two-Dimensional Chromatography

In some cases, the chemistry within a given column can be insufficient to separate some analytes. It is possible to direct a series of unresolved peaks onto a second column with different physico-chemical (Chemical classification) properties. Since the mechanism of retention on this new solid support is different from the first dimensional separation, it can be possible to separate compounds that are indistinguishable by one-dimensional chromatography. The sample is spotted at one corner of a square plate,developed, air-dried, then rotated by 90° and usually redeveloped in a second solvent system.

Pyrolysis Gas Chromatography

Pyrolysis gas chromatography mass spectrometry is a method of chemical analysis in which the sample is heated to decomposition to produce smaller molecules that are separated by gas chromatography and detected using mass spectrometry.

Pyrolysis is the thermal decomposition of materials in an inert atmosphere or a vacuum. The sample is put into direct contact with a platinum wire, or placed in a quartz sample tube, and rapidly heated to 600–1000 °C. Depending on the application even higher temperatures are used. Three different heating techniques are used in actual pyrolyzers: Isothermal furnace, inductive heating (Curie Point filament), and resistive heating using platinum filaments. Large molecules cleave at their weakest points and produce smaller, more volatile fragments. These fragments can be separated by gas chromatography. Pyrolysis GC chromatograms are typically complex because a wide range of different decomposition products is formed. The data can either be used as fingerprint to prove material identity or the GC/MS data is used to identify individual fragments to obtain structural information. To increase the volatility of polar fragments, various methylating reagents can be added to a sample before pyrolysis.

Besides the usage of dedicated pyrolyzers, pyrolysis GC of solid and liquid samples can be performed directly inside Programmable Temperature Vaporizer (PTV) injectors that provide quick heating (up to 30 °C/s) and high maximum temperatures of 600–650 °C. This is sufficient for some pyrolysis applications. The main advantage is that no dedicated instrument has to be purchased and pyrolysis can be performed as part of routine GC analysis. In this case quartz GC inlet liners have to be used. Quantitative data can be acquired, and good results of derivatization inside the PTV injector are published as well.

Applications

Countercurrent chromatography and related liquid-liquid separation techniques have been used on both industrial and laboratory scale to purify a wide variety of chemical substances. Separation realizations include proteins, DNA, antibiotics, vitamins, natural products, pharmaceuticals, metal ions, pesticides, enantiomers, polyaromatic hydrocarbons from environmental samples,active enzymes, and carbon nanotubes. Countercurrent chromatography is known for its high dynamic range of scalability: milligram to kilogram quantities purified chemical components may be obtained with this technique. It also has the advantage of accommodating chemically complex samples with undissolved particulates.

REFERENCES

1. Berthod, Alain; Maryutina, Tatyana; Spivakov, Boris; Shpigun, Oleg; Sutherland, Ian A. (2009). Countercurrent chromatography in analytical chemistry (IUPAC Technical Report). Pure and Applied Chemistry. 81 *(2).* doi:10.1351/PAC-REP-08-06-05. ISSN 1365-3075.
2. Ito, Y.; Bowman, RL (1970). Countercurrent Chromatography: Liquid-Liquid Partition Chromatography without Solid Support. Science. 167 *(3916): 281–283. Bibcode*:1970Sci...167..281I. doi:10.1126/science.167.3916.281. PMID 5409709.
3. High-speed countercurrent chromatography. Chemical analysis. Yoichiro Ito, Walter D. Conway (eds.). New York: J. Wiley. 1996. ISBN 978-0-471-63749-3.
4. Liu, Yang; Friesen, J. Brent; McApline, James B.; Pauli, Guido F. (2015). Solvent System Selection Strategies in Countercurrent Separation. Planta Medica.

81: 1582–1591. doi:10.1055/s-0035-1546246.

5. Ito, Yuko; Goto, Tomomi; Yamada, Sadaji; Matsumoto, Hiroshi; Oka, Hisao; Takahashi, Nobuyuki; Nakazawa, Hiroyuki; Nagase, Hisamitsu; Ito, Yoichiro (2006). Application of dual counter-current chromatography for rapid sample preparation of N-methylcarbamate pesticides in vegetable oil and citrus fruit. Journal of Chromatography A. 1108 *(1): 20–25. doi:10.1016/j.chroma.2005.12.070. Retrieved 2016-02-21.*
6. Tanimura, Takenori; Pisano, John J.; Ito, Yoichiro; Bowman, Robert L. (1970). Droplet Countercurrent Chromatography. Science. 169: 54–56. doi:10.1126/science.169.3940.54.
7. Foucault, Alain P. (1994). Centrifugal Partition Chromatography. Chromatographic Science Series, Vol. 68. CRC Press. ISBN 978-0-8247-9257-2.
8. Marchal, Luc; Legrand, Jack; Foucault, Alain (2003). Centrifugal partition chromatography: A survey of its history, and our recent advances in the field. The Chemical Record. 3 *(3): 133–143. doi:10.1002/tcr.10057.*
9. Ito, Yoichiro (2005). Golden rules and pitfalls in selecting optimum conditions for high-speed counter-current chromatography. Journal of Chromatography A. 1065 *(2): 145–168. doi:10.1016/j.chroma.2004.12.044.*
10. Randall, Merle; Longtin, Bruce (1938). SEPARATION PROCESSES General Method of Analysis. Industrial & Engineering Chemistry. 30 *(9): 1063–1067.* doi:10.1021/ie50345a028. Retrieved 2016-02-22.
11. Martin, A J P; Synge, R L M (1941). Separation of the higher monoamino-acids by counter-current liquid-liquid extraction: the amino-acid composition of wool. Biochemical Journal. 35 *(1-2): 91–121.* doi:10.1042/bj0350091. PMC 1265473. PMID 16747393.
12. Lyman C. Craig (1944). Identification of Small Amounts of Organic Compounds by Distribution Studies: II. Separation by Counter-current Distribution. Journal of Biological Chemistry. 155: 535–546.
13. Kimura, Yukio; Kitamura, Hisami; Hayashi, Kyozo (1982). A method for separating commercial colistin complex into new components: colistins pro-A, pro-B and pro-C.. The Journal of Antibiotics. 35 *(11): 1513–1520. doi:10.7164/antibiotics.35.1513. Retrieved 2016-02-22.*
14. Ian A. Sutherland (2007). Recent progress on the industrial scale-up of counter-current chromatography. Journal of Chromatography A. 1151: 6–13. doi:10.1016/j.chroma.2007.01.143.
15. DeAmicis, Carl; Edwards, Neil A.; Giles, Michael B.; Harris, Guy H.; Hewitson, Peter; Janaway, Lee; Ignatova, Svetlana (2011). Comparison of preparative reversed phase liquid chromatography and countercurrent chromatography for the kilogram scale purification of crude spinetoram insecticide. Journal of Chromatography A. 1218 *(36): 6122–6127. doi:10.1016/j.chroma.2011.06.073. Retrieved 2016-02-22.*
16. Hao Liang, Cuijuan Li, Qipeng Yuan and Frank Vriesekoop (2008). Application of High-Speed Countercurrent Chromatography for the Isolation of Sulforaphane

from Broccoli Seed Meal. J. Agric. Food Chem. 56: 7746–7749. doi:10.1021/jf801318v.

17. Conway, Walter D. (2011). Counter-current chromatography: Simple process and confusing terminology. Journal of Chromatography A. 1218 *(36): 6015–6023. doi:10.1016/j.chroma.2011.03.056. Retrieved 2016-02-22.*
18. Yoichiro Ito (2005). Golden rules and pitfalls in selecting optimum conditions for high-speed counter-current chromatography. Journal of Chromatography A. 1065: 145–168. doi:10.1016/j.chroma.2004.12.044.
19. Ito, Yoichiro (2014). Counter-current motion in counter-current chromatography. Journal of Chromatography A. 1372: 128–132. doi:10.1016/j.chroma.2014.09.033. Retrieved 2016-02-22.
20. Brown, Leslie; Luu, Trinh A. (2010). An introduction to discussions on liquid-liquid/counter current/centrifugal partition chromatography biphasic solvent selection methodologies and instrumentation for target compound preparation from complex matrices. Journal of Separation Science: 999–1003. doi:10.1002/jssc.200900814. Retrieved 2016-02-22.
21. Berthod, Alain (2014). Comments on Counter-current motion in counter-current chromatography by Yoichiro Ito. Journal of Chromatography A. 1372: 260–261. doi:10.1016/j.chroma.2014.10.103. Retrieved 2016-02-22.
22. Sutherland, Ian A. (2007). Recent progress on the industrial scale-up of counter-current chromatography. Journal of Chromatography A. 1151 *(1-2): 6–13. doi:10.1016/j.chroma.2007.01.143. Retrieved 2016-02-22.*
23. Ignatova, Svetlana; Wood, Philip; Hawes, David; Janaway, Lee; Keay, David; Sutherland, Ian (2007). Feasibility of scaling from pilot to process scale. Journal of Chromatography A. 1151 *(1-2): 20–24. doi:10.1016/j.chroma.2007.02.084. Retrieved 2016-02-22.*
24. Friesen, J. Brent; Pauli, Guido F. (2008). Performance Characteristics of Countercurrent Separation in Analysis of Natural Products of Agricultural Significance. Journal of Agricultural and Food Chemistry. 56 *(1): 19–28.* doi:10.1021/jf072415a. Retrieved 2016-02-25.
25. Brent Friesen, J.; Pauli, Guido F. (2005). G.U.E.S.S.—A Generally Useful Estimate of Solvent Systems for CCC. Journal of Liquid Chromatography & Related Technologies. 28 *(17): 2777–2806.* doi:10.1080/10826070500225234. Retrieved 2016-02-25.
26. Friesen, J. Brent; Pauli, Guido F. (2007). Rational development of solvent system families in counter-current chromatography. Journal of Chromatography A. 1151 *(1-2): 51–59. doi:10.1016/j.chroma.2007.01.126. Retrieved 2016-02-25.*
27. Friesen, J. Brent; Pauli, Guido F. (2009). GUESSmix-guided optimization of elution–extrusion counter-current separations. Journal of Chromatography A. 1216 *(19): 4225–4231. doi:10.1016/j.chroma.2008.12.053. Retrieved 2016-02-25.*
28. Berthod, A.; Carda-Broch, S. (2004). Use of the ionic liquid 1-butyl-3-methylimidazolium hexafluorophosphate in countercurrent chromatography. Analytical and Bioanalytical Chemistry. 380 *(1).* doi:10.1007/s00216-004-2717-8.

Retrieved 2016-02-25.

29. Berthod, A.; Ruiz-*Ángel*, M.J.; Carda-Broch, S. (2008). Ionic liquids in separation techniques. Journal of Chromatography A. 1184 *(1-2): 6–18. doi:10.1016/j.chroma.2007.11.109. Retrieved 2016-02-25.*

30. Friesen, J. Brent; McAlpine, James B.; Chen, Shao-Nong; Pauli, Guido F. (2015). Countercurrent Separation of Natural Products: An Update. Journal of Natural Products. 78 *(7): 1765–1796.* doi:10.1021/np501065h. Retrieved 2016-02-21.

31. Pauli, Guido F.; Pro, Samuel M.; Friesen, J. Brent (2008). Countercurrent Separation of Natural Products. Journal of Natural Products. 71 *(8): 1489–1508.* doi:10.1021/np800144q.

32. Tanimura, T.; Pisano, J. J.; Ito, Y.; Bowman, R. L. (1970). Droplet Countercurrent Chromatography. Science. 169 *(3940): 54–56. doi:10.1126/science.169.3940.54. Retrieved 2016-02-22.*

33. Hostettmann, Kurt; Hostettmann-Kaldas, Maryse; Sticher, Otto (1979). Application of droplet counter-current chromatography to the isolation of natural products. Journal of Chromatography A. 186: 529–534. doi:10.1016/S0021-9673(00)95273-7. Retrieved 2016-02-22.

34. Ito, Y.; Weinstein, M.; Aoki, I.; Harada, R.; Kimura, E.; Nunogaki, K. (1966). The Coil Planet Centrifuge. Nature. 212 *(5066): 985–987.* doi:10.1038/212985a0. Retrieved 2016-02-22.

35. Ito, Y.; Bowman, R. L. (1971). Countercurrent Chromatography with Flow-Through Coil Planet Centrifuge. Science. 173 *(3995): 420–422. doi:10.1126/science.173.3995.420. Retrieved 2016-02-22.*

36. Ito, Y; Suaudeau, J; Bowman, R. (1975). New flow-through centrifuge without rotating seals applied to plasmapheresis. Science. 189 *(4207): 999–1000. doi:10.1126/science.1220011. Retrieved 2016-02-22.*

37. Ito, Yoichiro (1981). Countercurrent chromatography. Journal of Biochemical and Biophysical Methods. 5 *(2): 105–129.* doi:10.1016/0165-022X(81)90011-7. Retrieved 2016-02-22.

38. Friesen, J.B.; Pauli, G.F. (2009). Binary concepts and standardization in counter-current separation technology. Journal of Chromatography A. 1216 *(19): 4237–4244. doi:10.1016/j.chroma.2009.01.048. Retrieved 2016-02-22.*

39. Ito, Yoichiro (2005). Origin and Evolution of the Coil Planet Centrifuge: A Personal Reflection of My 40 Years of CCC Research and Development. Separation & Purification Reviews. 34 *(2): 131–154.* doi:10.1080/15422110500322883. Retrieved 2016-02-25.

40. Ito, Yoichiro; Sandlin, Jesse; Bowers, William G. (1982). High-speed preparative counter-current chromatography with a coil planet centrifuge. Journal of Chromatography A. 244 *(2): 247–258.* doi:10.1016/S0021-9673(00)85688-5. Retrieved 2016-02-22.

41. Fales, Henry M.; Pannell, Lewis K.; Sokoloski, Edward A.; Carmeci, Peter. (1985). Separation of Methyl Violet 2B by high-speed countercurrent chromatography

and identification by californium-252 plasma desorption mass spectrometry. Analytical Chemistry. 57 *(1): 376–378.* doi:10.1021/ac00279a089. Retrieved 2016-02-22.

42. Conway, Walter D. (1990). Countercurrent Chromatography: Apparatus, Theory and Applications. New York: VCH Publishers. ISBN 978-0-89573-331-3.
43. Eulogy for Ed Chou. Journal of Liquid Chromatography & Related Technologies. 28 *(12-13): 1789–1791. 2005.* doi:10.1081/JLC-200063429. Retrieved 2016-02-22.
44. Sutherland, I. A.; Brown, L.; Forbes, S.; Games, G.; Hawes, D.; Hostettmann, K.; McKerrell, E. H.; Marston, A.; Wheatley, D.; Wood, P. (1998). Countercurrent Chromatography (CCC) and its Versatile Application as an Industrial Purification & Production Process. Journal of Liquid Chromatography & Related Technologies. 21 *(3): 279–298.* doi:10.1080/10826079808000491. Retrieved 2016-02-22.
45. Berthod, A. (2002). Countercurrent Chromatography: The support-free liquid stationary phase. Wilson & Wilson`s Comprehensive Analytical Chemistry Vol. 38. Boston: Elsevier Science Ltd. pp. 1–397. ISBN 978-0-444-50737-2.
46. Conway, Walter D.; Petroski, Richard J. (1995). Modern Countercurrent Chromatography. ACS Symposium Series #593. ACS Publications. doi:10.1021/bk-1995-0593. ISBN 978-0-8412-3167-2.
47. Ito, Yoichiro; Conway, Walter D. (1995). High-speed Countercurrent Chromatography. Chemical Analysis: A Series of Monographs on Analytical Chemistry and Its Applications (Book 198). New York: John Wiley and Sons. ISBN 978-0-471-63749-3.
48. Mandava, N. Bhushan; Ito, Yoichiro (1988). Countercurrent Chromatography: Theory and Practice. Chromatographic Science Series, Vol. 44. New York: Marcel Dekker Inc. ISBN 978-0-8247-7815-6.
49. Menet, Jean-Michel; Thiebaut, Didier (1999). Countercurrent Chromatography. Chromatographic Science Series Vol 82. New York: CRC Press. ISBN 978-0-8247-9992-2.
50. Hacer Guzlek; et al. (2009). Performance comparison using the GUESS mixture to evaluate counter-current chromatography instruments. Journal of Chromatography A. 1216 *(19): 4181–4186. doi:10.1016/j.chroma.2009.02.091. PMID 19344911.*
51. ^ Jump up to:[a] [b] Sumner, Neil (2011). Developing counter current chromatography to meet the needs of pharmaceutical discovery. Journal of Chromatography A. 1218 *(36): 6107–6113. doi:10.1016/j.chroma.2011.05.001. Retrieved 2016-02-21.*
52. Wataru Murayama, Tetsuya Kobayashi, Yasutaka Kosuge, Hideki Yano1, Yoshiaki Nunogaki, and Kanichi Nunogaki (1982). A new centrifugal counter-current chromatograph and its application. Journal of Chromatography A. 239: 643–649. doi:10.1016/S0021-9673(00)82022-1.
53. Margraff, Rodolphe; Intes, Olivier; Renault, Jean-Hugues; Garret, Pierre (2005). “Partitron 25, a Multi-Purpose Industrial Centrifugal Partition Chromatograph: Rotor Design and Preliminary Results on Efficiency and Stationary Phase Retention”. Journal of Liquid Chromatography & Related Technologies. 28 *(12-13): 1893–1902.* doi:10.1081/JLC-200063539. Retrieved 2016-02-22.

54. Hamzaoui, Mahmoud; Hubert, Jane; Hadj-Salem, Jamila; Richard, Bernard; Harakat, Dominique; Marchal, Luc; Foucault, Alain; Lavaud, Catherine; Renault, Jean-Hugues (2011). Intensified extraction of ionized natural products by ion pair centrifugal partition extraction. Journal of Chromatography A. 1218 *(31): 5254–5262. doi:10.1016/j.chroma.2011.06.018. Retrieved 2016-02-23.*

55. Sutherland, Ian; Hewitson, Peter; Siebers, Rick; van den Heuvel, Remco; Arbenz, Lillian; Kinkel, Joachim; Fisher, Derek (2011). Scale-up of protein purifications using aqueous two-phase systems: Comparing multilayer toroidal coil chromatography with centrifugal partition chromatography. Journal of Chromatography A. 1218 *(32): 5527–5530. doi:10.1016/j.chroma.2011.04.013. Retrieved 2016-02-22.*

56. Faure, Karine; Bouju, Elodie; Suchet, Pauline; Berthod, Alain (2013). Use of Limonene in Countercurrent Chromatography: A Green Alkane Substitute. Analytical Chemistry. 85 *(9): 4644–4650.* doi:10.1021/ac4002854. Retrieved 2016-02-23.

57. Marchal, L.; Foucault, A.; Patissier, G.; Rosant, J.M.; Legrand, J. (2000). Influence of flow patterns on chromatographic efficiency in centrifugal partition chromatography. Journal of Chromatography A. 869 *(1-2): 339–352.* doi:10.1016/S0021-9673(99)01184-X. Retrieved 2016-02-22.

58. Ingkaninan, K.; Hazekamp, A.; Hoek, A. C.; Balconi, S.; Verpoorte, R. (2000). Application of Centrifugal Partition Chromatography in a General Separation and Dereplication Procedure for Plant Extracts. Journal of Liquid Chromatography & Related Technologies. 23 *(14): 2195–2208.* doi:10.1081/JLC-100100481. Retrieved 2016-02-24.

59. Berthod, Alain; Ruiz-Angel, Maria Jose; Carda-Broch, Samuel (2003). Elution–Extrusion Countercurrent Chromatography. Use of the Liquid Nature of the Stationary Phase To Extend the Hydrophobicity Window. Analytical Chemistry. 75 *(21): 5886–5894.* doi:10.1021/ac030208d. Retrieved 2016-02-24.

60. Wu, Dingfang; Cao, Xiaoji; Wu, Shihua (2012). Overlapping elution–extrusion counter-current chromatography: A novel method for efficient purification of natural cytotoxic andrographolides from Andrographis paniculata. Journal of Chromatography A. 1223: 53–63. doi:10.1016/j.chroma.2011.12.036. Retrieved 2016-02-24.

61. Cao, Xueli; Wang, Qiaoe; Li, Yan; Bai, Ge; Ren, Hong; Xu, Chunming; Ito, Yoichiro (2011). Isolation and purification of series bioactive components from Hypericum perforatum L. by counter-current chromatography. Journal of Chromatography B. 879 *(7-8): 480–488. doi:10.1016/j.jchromb.2011.01.007. Retrieved 2016-02-23.*

62. Ignatova, Svetlana; Sumner, Neil; Colclough, Nicola; Sutherland, Ian (2011). Gradient elution in counter-current chromatography: A new layout for an old path. Journal of Chromatography A. 1218 *(36): 6053–6060. doi:10.1016/j.chroma.2011.02.052. Retrieved 2016-02-23.*

63. Agnely, M; Thiébaut, D (1997). Dual-mode high-speed counter-current

chromatography: retention, resolution and examples. Journal of Chromatography A. 790 *(1-2): 17–30.* doi:10.1016/S0021-9673(97)00742-5. Retrieved 2016-02-23.

64. Ito, Yoichiro (1985). Foam Countercurrent Chromatography Based on Dual Counter-Current System. Journal of Liquid Chromatography. 8 *(12): 2131–2152.* doi:10.1080/01483918508074122. Retrieved 2016-02-24.

65. Lee, Y.-W.; Cook, C. E.; Ito, Y. (1988). Dual Countercurrent Chromatography. Journal of Liquid Chromatography. 11 *(1): 37–53.* doi:10.1080/01483919808068313. Retrieved 2016-02-24.

66. van den Heuvel, Remco; Sutherland, Ian (2007). Observations of established dual flow in a spiral dual-flow counter-current chromatography coil. Journal of Chromatography A. 1151 *(1-2): 99–102. doi:10.1016/j.chroma.2007.01.099. Retrieved 2016-02-24.*

67. Ignatova, Svetlana; Hewitson, Peter; Mathews, Ben; Sutherland, Ian (2011). Evaluation of dual flow counter-current chromatography and intermittent counter-current extraction. Journal of Chromatography A. 1218 *(36): 6102–6106. doi:10.1016/j.chroma.2011.02.032. Retrieved 2016-02-24.*

68. Sidana, Jasmeen; Joshi, Lokesh Kumar (2013). Recycle HPLC: A Powerful Tool for the Purification of Natural Products. Chromatography Research International. 2013: 1–7. doi:10.1155/2013/509812. Retrieved 2016-02-24.

69. Du, Q.-Z.; Ke, C.-Q.; Ito, Y. (1998). Recycling High-Speed Countercurrent Chromatography for Separation of Taxol and Cephalomannine. Journal of Liquid Chromatography & Related Technologies. 21 *(1-2): 157–162.* doi:10.1080/10826079808001944. Retrieved 2016-02-24.

70. Meng, Jie; Yang, Zhi; Liang, Junling; Guo, Mengze; Wu, Shihua (2014). Multi-channel recycling counter-current chromatography for natural product isolation: Tanshinones as examples. Journal of Chromatography A. 1327: 27–38. doi:10.1016/j.chroma.2013.12.069. Retrieved 2016-02-24.

71. Englert, Michael; Brown, Leslie; Vetter, Walter (2015). Heart-Cut Two-Dimensional Countercurrent Chromatography with a Single Instrument. Analytical Chemistry. 87 *(20): 10172–10177. doi:10.1021/acs.analchem.5b02859. Retrieved 2016-02-24.*

72. Qiu, Feng; Friesen, J. Brent; McAlpine, James B.; Pauli, Guido F. (2012). Design of countercurrent separation of Ginkgo biloba terpene lactones by nuclear magnetic resonance. Journal of Chromatography A. 1242: 26–34. doi:10.1016/j.chroma.2012.03.081. Retrieved 2016-02-24.

73. Maciuk, Alexandre; Renault, Jean-Hugues; Margraff, Rodolphe; Trébuchet, Philippe; Zèches-Hanrot, Monique; Nuzillard, Jean-Marc (2004). Anion-Exchange Displacement Centrifugal Partition Chromatography. Analytical Chemistry. 76 *(21): 6179–6186.* doi:10.1021/ac049499w. Retrieved 2016-02-24.

74. Toribio, Alix; Nuzillard, Jean-Marc; Renault, Jean-Hugues (2007). Strong ion-exchange centrifugal partition chromatography as an efficient method for the large-scale purification of glucosinolates. Journal of Chromatography A. 1170 *(1-2): 44–51. doi:10.1016/j.chroma.2007.09.004. Retrieved 2016-02-24.*

75. Ito, Yoichiro (2013). pH-zone-refining counter-current chromatography: Origin, mechanism, procedure and applications. Journal of Chromatography A. 1271 *(1): 71–85. doi:10.1016/j.chroma.2012.11.024. Retrieved 2016-02-25.*

76. Fang, Lei; Zhou, Jie; Lin, YunLiang; Wang, Xiao; Sun, Qinglei; Li, Jia-Lian; Huang, Luqi (2013). Large-scale separation of alkaloids from Gelsemium elegans *by pH-zone-refining counter-current chromatography with a new solvent system screening method". Journal of Chromatography A.* 1307: 80–85. doi:10.1016/j.chroma.2013.07.069. Retrieved 2016-02-25.

77. Englert, Michael; Vetter, Walter (2015). Overcoming the equivalent-chain-length rule with pH-zone-refining countercurrent chromatography for the preparative separation of fatty acids. Analytical and Bioanalytical Chemistry. 407 *(18): 5503–5511.* doi:10.1007/s00216-015-8723-1. Retrieved 2016-02-25.

78. Mekaoui, Nazim; Faure, Karine; Berthod, Alain (2012). Advances in Countercurrent Chromatography for Protein Separations. Bioanalysis. 4: 833–844. doi:10.4155/bio.12.27.

79. Kendall, D.; Booth, A. J.; Levy, M.S.; Lye, G. J. (2001). Separation of supercoiled and open-circular plasmid DNA by liquid-liquid counter-current chromatography. Biotechnology Letters. 23: 613–619. doi:10.1023/A:1010362812469.

80. McAlpine, James B.; Friesen, J. Brent; Pauli, Guido F. (2012). Separation of Natural Products by Countercurrent Chromatography. In Satyajit D. Sarker, Lutfun Nahar (eds.). Natural Products Isolation. 864. Totowa, NJ: Humana Press. pp. 221–254. Retrieved 2016-02-21.

81. Kurumaya, Katsuyuki; Sakamoto, Tetsuto; Okada, Yoshihito; Kajiwara, Masahiro (1988). Application of droplet counter-current chromatography to the isolation of vitamin B12. Journal of Chromatography A. 435: 235–240. doi:10.1016/S0021-9673(01)82181-6. Retrieved 2016-02-21.

82. Araki, T.; Okazawa, T.; Kubo, Y.; Ando, H.; Asai, H. (1988). Separation of Lighter Rare Earth Metal Ions by Centrifugal Countercurrent Type Chromatography with Di-(2-ethylhexyl)phosphoric Acid. Journal of Liquid Chromatography. 11 *(1): 267–281.* doi:10.1080/01483919808068328. Retrieved 2016-02-16.

83. Ito, Yuko; Goto, Tomomi; Yamada, Sadaji A.; Ohno, Tsutomu; Matsumoto, Hiroshi; Oka, Hisao; Ito, Yoichiro (2008). Rapid determination of carbamate pesticides in food using dual counter-current chromatography directly interfaced with mass spectrometry. Journal of Chromatography A. 1187: 53–57. doi:10.1016/j.chroma.2008.01.078.

84. Chiral recognition in separation methods: mechanisms and applications. Alain Berthod (ed.). Heidelberg ; New York: Springer. 2010.

85. Cao, Xueli; Yang, Chunlei; Pei, Hairun; Li, Xinghong; Xu, Xiaobai; Ito, Yoichiro (2012). Application of counter-current chromatography as a new pretreatment method for the determination of polycyclic aromatic hydrocarbons in environmental water. Journal of Separation Science. 35 *(4): 596–601. doi:10.1002/jssc.201100852. Retrieved 2016-02-21.*

86. Baldermann, Susanne; Mulyadi, Andriati N.; Yang, Ziyin; Murata, Ariaka; Fleischmann, Peter; Winterhalter, Peter; Knight, Martha; Finn, Thomas M.; Watanabe, Naoharu (2011). Application of centrifugal precipitation chromatography and high-speed counter-current chromatography equipped with a spiral tubing support rotor for the isolation and partial characterization of carotenoid cleavage-like enzymes in Enteromorpha compressa. Journal of Separation Science. 34 *(19): 2759–2764. doi:10.1002/jssc.201100508. Retrieved 2016-02-21.*

87. Zhang, Min; Khripin, Constantine Y.; Fagan, Jeffrey A.; McPhie, Peter; Ito, Yoichiro; Zheng, Ming (2014). Single-Step Total Fractionation of Single-Wall Carbon Nanotubes by Countercurrent Chromatography. Analytical Chemistry. 86: 3980–3984. doi:10.1021/ac5003189.

88. Sutherland, Ian A. (2007). Recent progress on the industrial scale-up of counter-current chromatography. Journal of Chromatography A. 1151: 6–13. doi:10.1016/j.chroma.2007.01.143.

CHAPTER 8

Chromatography Detectors

Chapter Contents

- **Introduction**
- **Destructive Detect**
- **Non-Destructictors**
- **Flame Ionization Detector**
- **Advantages and Disadvantages**
- **Nitrogen–Phosphorus Detector**
- **Evaporative Light Scattering Detector (Elsd)**
- **Mass Spectrometry**

A detector can be compared to a gate watchman who verifies the visitors before permitting them entry inside a building. The chromatographic detector is capable of establishing both the identity and concentration of eluting components in the mobile phase stream. A broad range of detectors is available to meet different sample requirements. Specific detectors respond to a particular compound only and the response is independent of mobile phase composition. On the other hand the response of bulk property detectors is dependent on collective changes in composition of sample and mobile phase.

INTRODUCTION

A **chromatography detector** is a device used in gas chromatography (GC) or liquid chromatography (LC) to detect components of the mixture being eluted off the chromatography column. There are two general types of detectors: destructive and non-destructive. The destructive detectors perform continuous transformation of the column effluent (burning, evaporation or mixing with reagents) with subsequent measurement of some physical property of the resulting material (plasma, aerosol or reaction mixture). The non-destructive detectors are directly measuring some property of the column eluent (for example UV absorption) and thus affords for the further analysis recovery.

DESTRUCTIVE DETECT

- Flame ionization detector (FID)
- Aerosol-based detector (NQA)
- Flame photometric detector (FPD)
- Atomic-emission detector (AED)
- Nitrogen Phosphorus Detector (NPD)
- Evaporative light scattering detector (ELSD)
- Mass spectrometer (MS)
- Electrolytic Conductivity detector (ELCD)
- sumon detector (SMSD)
- Mira detector (MD)

NON-DESTRUCTICTORS

- UV detectors, fixed or variable wavelength, which includes diode array detector (DAD or PDA). The UV absorption of the effluent is continuously measured at single or multiple wavelengths. These are by far most popular detec for LC.
- Thermal conductivity detector, (TCD). Measures the thermal conductivity of the eluent. Only used in GC.
- Fluorescence detector. Irradiates the effluent with a light of set wavelength and measure the fluorescence of the effluent at a single or multiple wavelength. Used only in LC
- Electron capture detector, (ECD). The most sensitive detector known. Allows for the detection of organic molecules containing halogen, nitro groups etc.
- Conductty monitor. Continuously measures the conductivity of the effluent. Used only in LC when conductive eluents (water or alcohols) are used.
- Photoionization detector, (PID). Measures the increase in conductivity achieved by ionizing the effluent gas with UV radiation.
- Refractive index detector (RI or RID). Continuously measures the refractive index of the effluent. Used only in LC. The lowest sensitivity of all detectors. Often used in size exclusion chromatography for polymer analysis.
- Radio flow detector. Measures radioactivity of the effluent. This detector can be destructive if a scintillation cocktail is continuously added to the effluent.
- Chiral detector continuously measures the optical angle of rotation of the effluent. It is used only in LC when chiral compounds are being analyzed.

FLAME IONIZATION DETECTOR

A **flame ionization detector (FID)** is a scientific instrument that measures the concentration of organic species in a gas stream. It is frequently used as a detector in gas chromatography. Standalone FIDs can also be used in applications such as landfill gas monitoring, fugitive emissions monitoring and internal combustion engine emissi measurement in stationary or portable instru

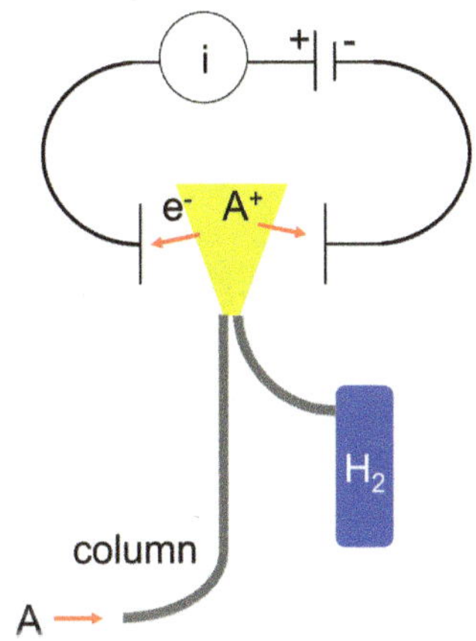

Schematic of a flame ionization detector for gas chromatography.

History

The first flame ionization detectors were developed simultaneously and independently in 1957 by McWilliam and Dewar at Imperial Chemical Industries of Australia and New Zealand (ICIANZ, see Orica#History) Central Research Laboratory, Ascot Vale, Melbouralia. and by Harley and Pretorius at the University of Pretoria in Pretoriauth Africa.

In 1959, Perkin Elmer Corp. included a flame ionization detector in its Varactometer

Operrinciple

The operation of the FID is based on the detection of ions formed during combustion of organic compounds in a hydrogen flame. The generation of these ions is proportional to the concentration of organic species in the sample gas stream.

FID measurements are usually reported as "as methane", meaning as the quantity of methane which would produce the same response. Hydrocarbons generally have molar response factors that are equal to the number of carbon atoms in their molecule, while oxygenates and other species that contain heteroatoms tend to have a lower response factor. Carbon monoxide and carbon dioxide are not detectable by FID.

FID measurements are often labelledtotaydrocarbons" or "total hydrocarbon content" (THC), although a more accurate name would betotal volatile hydrocarbon tent" (TVHC), as hydrocarbons which have condensed out are not detected, even though they are important for e.g. safety when handling compressed oxygen.

ADVANTAGES AND DISADVANTAGES

Flame ionization detectors are used very widely in gas chromatography because of a number of advantages.

- Cost: Flame ionization detectors are relatively inexpensive to acquire and operate.
- Low maintenance requirements: Apart from cleaning or replacing the FID jet, these detectors require little maintenance.
- Rugged construction: FIDs are relatively resistant to misuse.
- Linearity and detection ranges: FIDs can measure organic substance concentration at very low and very high levels, having a linear response range of 10^7**dvantages**

Flame ionization detectors cannot detect inorganic substances and some highly oxygenated or functionalized species like infrared and laser technology can. In some systems, CO and CO_2 can be detected in the FID using a methanizer, which is a bed of Ni catalyst that reduces CO and CO_2 to methane, which can be in turn detected by the FID. The methanizer is limited by its inability to reduce compounds other than CO and CO_2 and its tendency to be poisoned by a number of chemicals commonly found in GC effluents.

Another important disadvantage is that the FID flame oxidizes all oxidizable compounds that pass through it; all hydrocarbons and oxygenates are oxidized to carbon dioxide and water and other heteroatoms are oxidized according to thermodynamics. For this reason, FIDs tend to be the last in a detector train and also cannot be used for preparatory work.

Alternative Solution

An improvement to the methanizer exists in the form of a sequential reactor that oxidizes compounds before reducing them to methane. This method can be used to improve the response of the FID and allow for the detection of many more carbon-contang compounds. The complete conversion of compounds to methane and the now equivalent response in the detector also eliminates the need for calibrations and standards because response factors are all equivalent to those of methane. This allows for the rapid analysis of complex mixtures that contain molecules where standards are not available. The sequential reactor is sold commercially as the Polyarc reactor, available online from the Activated Rh Company

Operation

D Schematic:[11] A) Capillary tube; B) Platinum jet; C) Hydrogen; D) Air; E) Flame; F) Ions; G) Collector; H) Coaxial cable to Analog to Digital converter; J) Gas outlet.

To detect these ions, two electrodes are used to provide a potential difference. The positive electrode doubles as the nozzle head where the flame is produced. The other, negative electrode is positioned above the flame. When first designed, the negative electrode was either tear-drop shaped or angular piece of platinum. Today, the design has been modified into a tubular electrode, commonly referred to as a collector plate. The ions thus are attracted to the collector plate and upon hitting the plate, induce a current. This current is measured with a high-impedance picoammeter and fed into an integrator. The manner in which the final data is displayed is based on the computer and software. In general, a graph is displayed that has time on the x-axis and total ion on the y-axis.

The current measured corresponds roughly to the proportion of reduced carbon atoms in the flame. Specifically how the ions are produced is not necessarily understood, but the response of the detector is determined by the number of carbon atoms (ions) hitting the detector per unit time. This makes the detector sensitive to the mass rather than the concentration, which is useful because the response of the detector is not greatly affected by changes in the carrier gas flow rate.

Description of a Generic Detector

The design of the flame ionization detector varies from manufacturer to manufacturer, but the principles are the same. Most commonly, the FID is attached to a gas chromatography system.

The eluent exits the GC column (A) and enters the FID detector's oven (B). The oven is needed to make sure that as soon as the eluent exits the column, it does not come out of the gaseous phase and deposit on the interface between the column and FID. This deposition would result in loss of eluent and errors in detection. As the eluent travels up the FID, it is first mixed with the hydrogen fuel (C) and then with the oxidant (D). The eluent/fuel/oxidant mixture continues to travel up to the nozzle head where a positive bias voltage exists (E). This positive bias helps to repel the reduced carbon ions created by the flame (F) pyrolyzing the eluent. The ions are repelled up toward the collector plates (G) which are connected to a very sensitive ammeter, which detects the ions hitting the plates, then feeds that signal (H) to an amplifier, integrator, and display system. The products of the flame are finally vented out of the detector through the exhaust port (J

NITROGEN–PHOSPHORUS DETECTOR

The **nitrogen–phosphorus detector** (**NPD**) is also known as **thermionic specific detector** (**TSD**) is a type of detector commonly used with gas chromatography, in which thermal energy is used to ionize an analyte. It is a type of flame thermionic detector (FTD), the other being the alkali flame-ionization detector (AFID also known as AFD). With this method, nitrogen and phosphorus can be selectively detected with a sensitivity that is 10^4 times greater than that for carbon. NP-Mode: A concentration of hydrogen gas is used such that it is just below the minimum required for ignition. A rubidium or cesium bead, which is mounted over the nozzle, ignites the hydrogen (by acting catalytically), and forms a cold plasma. Excitation of the alkali metal results in ejection of electrons, which in turn are detected as a current flow between an anode and cathode in the chamber. As nitrogen or phosphorus analytes exit the column, they cause a reduction in the work function of the metal bead, resulting in an increase in current. Since the alkali metal bead is consumed over time, it must be replaced regularly .

EVAPORATIVE LIGHT SCATTERING DETECTOR (ELSD)

An **evaporative light scattering detector** (ELSD) is a detector used in conjunction with high-performance liquid chromatography (HPLC). It is commonly used for analysis of compounds where UV detection might be a restriction and therefore compounds does not very efficient absorb UV radiation, such as sugars, antivirals, antibiotics, lipids, phospholipids, terpenond alcohols. ELSDs fall under the category of general-purpose detectors, similar to refractive indtectors (RI).

Principles of Operation

ELSDs analyze solvent after elution from HPLC. As the eluent passes from an HPLC, it is mixed with an inert carrier gas and forced through a nebulizer, which separates the liquid into minute aerosolized droplets. These droplets then pass into a heated drift tube, where the mobile phase solvent is evaporated off. As the mobile phase evaporates, the droplets become smaller and smaller until all that is left is minute particles of dried analyte. These particles are pushed through the drift tube by the carrier gas to the detection region. In this region, a beam of light crosses the column of analyte and the scattering of light is measured by a photiplier tube.

MASS SPECTROMETRY

Mass spectrometry (**MS**) is an analytical technique that ionizes chemical species and sorts the ions based on their mass-to-charge ratio. In simpler terms, a mass spectrum measures the masses within a sample. Mass spectrometry is used in many different fields and is applied to pure samples as well as complex mixtures.

SIMS mass spectrometer, model IMS 3f.

A mass spectrum is a plot of the ion signal as a function of the mass-to-charge ratio. These spectra are used to determine the elemental or isotopic signature of a sample, the masses of particles and of molecules, and to elucidate the chemical structures of molecules, such as peptides and other chemical compounds.

In a typical MS procedure, a sample, which may be solid, liquid, or gas, is ionized, for example by bombarding it with electrons. This may cause some of the sample's molecules to break into charged fragments. These ions are then separated according to their mass-to-charge ratio, typically by accelerating them and subjecting them to an electric or magnetic field: ions of the same mass-to-charge ratio will undergo the same amo of deflection. The ions are detected by a mechanism capable of detecting charged particles, such as an electron multiplier. Results are displayed as spectra of the relative abundance of detected ions as a function of the mass-to-charge ratio. The atoms or molecules in the sample can be identified by correlating known masses to the identified masses or through a characteristic fragm14Exters spectrometry.

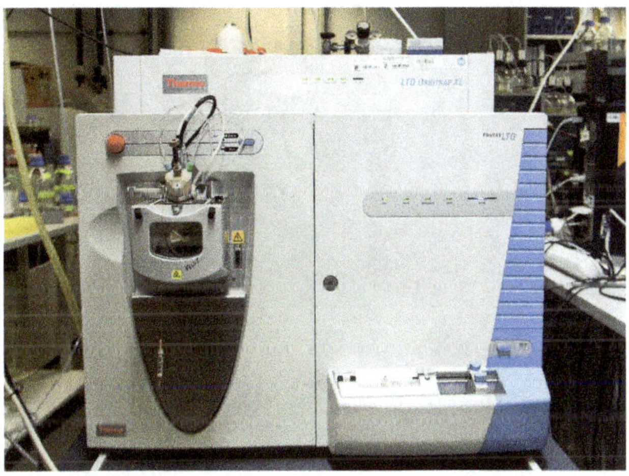

Orbitrap mass spectrometer.

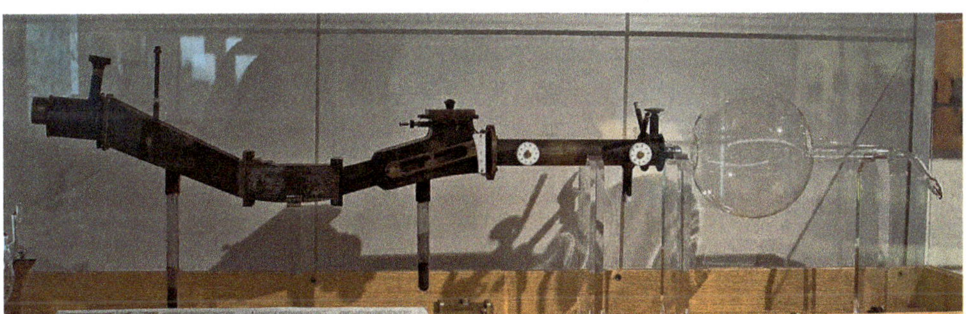

Replica of J.J. Thompson's third mass spectrometer.

In 1886, Eugen Goldstein observed rays in gas discharges under low pressure that traveled away from the anode and through channels in a perforated cathode, opposite to the direction of negatively charged cathode rays (which travel from cathode to anode). Goldstein called these positively charged anode rays "Kanalstrahlen"; the standard translation of this term into English iscanal rays". Wilhelm Wien found that strong electric or magnetic fields deflected the canal rays and, in 1899, constructed a device with parallel electric and magnetic fields that separated the positive rays according to their charge-to-mass ratio (*Q/m*). Wien found that the charge-to-mass ratio depended on the nature of the gas in the discharge tube. English scientist J.J. Thomson later improved on the work of Wien by reducing the pressure to create the mas spectrograph.

Calutron mass spectrometers were used in the Manhattan Project for uranium enrichment.

The word *spectrograph* had become part of the international scientific vary by 1884. Early *spectrometry* devices that measured the mass-to-charge ratio of ions were called mass spectrographs which consisted of instruments that recorded a spectrum of mass values on a paphic plate. A *mass spectroscope* is similar to a *mass spectrograph* except that the beam of ions is directed onto hosphor screen.A mass spectroscope configuration was used in early instruments when it was desired that the effects of adjustments be quickly observed. Once the instrument was properly adjusted, a photographic plate was inserted and exposed. The term mass spectroscope continued to be used even though the direct illumination of a phosphor screen was replaced by indirect measurements witn oscilloscope. The use of the term *mass spectroscopy* is now discouraged due to the possibility of confusion with lpectroscopy.Mass spectrometry is often abbreviated as *mass-spe* simply as *MS.*

Modern techniques of mass spectrometry were devised by Arthur Jeffrey Dempster and F.W. Aston in 1918 and 1919 respectively.

Sector mass spectrometers known as calutrons were used for separating the isotopes of uranium developed by Ernest O. Lawrence during the hattan Project. Calutron mass spectrometers were used for uranium enrichment at the Oak Ridge, Tennessee Y-12 plant established during World War II.

In 1989, half of the Nobel Prize in Physics was awarded to Hans Dehmelt and Wolfgang Paul for the development of the ion trap technique in the 1950s and 1960s.

In 2002, the Nobel Prize in Chemistry was awarded to John Bennett Fenn for the development of electrospray ionization (ESI) and Koichi Tanaka for the development of soft laser desorption (SLD) and their application to the ionization of biological macromolecules, ally proteins.[10]

Parts of a Mass Spectrometer

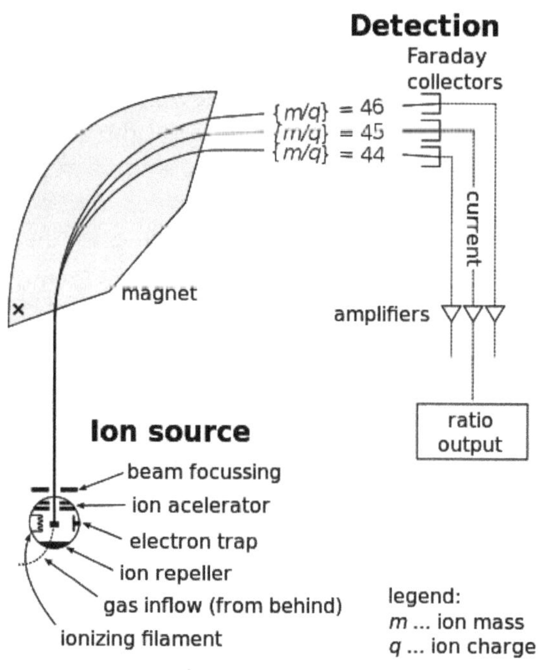

Schematics of a simple mass spectrometer with sector type mass analyzer. This one is for the measurement of carbon dioxide isotope ratios (IRMS) as in the carbon-13 urea breath test.

A mass spectrometer consists of three components: an ion source, a mass analyzer, and a detector. The ionizer converts a portion of the sample into ions. There is a wide variety of ionization techniques, depending on the phase (solid, liquid, gas) of the sample and the efficiency of various ionization mechanisms for the unknown species. An extraction system removes ions from the sample, which are then targeted through the mass analyzer and onto the *detector*. The differences in masses of the fragments allows the mass analyzer to sort the ions by their mass-to-charge ratio. The detector measures the value of an indicator quantity and thus provides data for calculating the abundances of each ion present. Some detectors also give spatial information, e.g., a multichannel plate.

Theoretical example

The following example describes the operation of a spectrometer mass analyzer, which is of the sector type. (Other analyzer types are treated below.) Consider a sample of sodium chloride (table salt). In the ion source, the sample is vaporized (turned into gas) and ionized (transformed into electrically charged particles) into sodium (Na^+) and chloride (Cl^-) ions. Sodium atoms and ions are monoisotopic, with a mass of about 23 u. Chloride atoms and ions come in two isotopes with masses of approximately 35 u (at a natural abundance of about 75 percent) and approximately 37 u (at a natural abundance of about 25 percent). The analyzer part of the spectrometer contains electric and magnetic fields, which exert forces on ions traveling through these fields. The speed of a charged particle may be increased or decreased while passing through the electric field, and its direction may be altered by the magnetic field. The magnitude of the deflection of the moving ion's trajectory depends on its mass-to-charge ratio. Lighter ions get deflected by the magnetic force more than heavier ions (based on Newton`s second law of motion, $F = ma$). The streams of sorted ions pass from the analyzer to the detector, which records the relative abundance of each ion type. This information is used to determine the chemical element composition of the original sample (i.e. that both sodium and chlorine are present in the sample) and the isotopic composition of its constituents (the ratio of ^{35}Cl to ^{37}Cl)

Creating ions

Surface ionization source at the Argonne National Laboratory linear accelerator.

The ion source is the part of the mass spectrometer that ionizes the material under analysis (the analyte). The ions are then transported by magnetic or electric fields to the mass analyzer.

Techniques for ionization have been key to determining what types of samples can be analyzed by mass spectrometry. Electron ionization and chemical ionization are used for gases and vapors. In chemical ionization sources, the analyte is ionized by chemical ion-molecule reactions during collisions in the source. Two techniques often used with liquid and solid biological samples include electrospray ionization (ited by John Fenn[11]) and matrix-assisted laser desorption/ionization (MALDI, initially developed as a similar technique "Soft Laser DesorptionD)" by K. Tanaka[12] for which a Nobel Prize was awarded and as MALDI by M. Karnd F. Hillenkamp[13]).

Hard Ionization and Soft Ionization

Quadrupole mass spectrometer and electrospray ion source used for Fenn's early work.

In mass spectrometry, ionization refers to the production of gas phase ions suitable for resolution in the mass analyser or mass filter. Ionization occurs in the ion source. There are several ion sources available; each has advantages and disadvantages for particular applications. For example, electron ionization (EI) gives a high degree of fragmentation, yielding highly detailed mass spectra which when skilfully analysed can provide important information for structural elucidation/characterisation and facilitate identification of unknown compounds by comparison to mass spectral libraries obtained under identical operating conditions. However, EI is not suitable for coupling to HPLC, i.e. LC-MS, since at atmospheric pressure, the filaments used to generate electrons burn out rapidly. Thus EI is coupled predominantly with GC, i.e. GC-MS, where the entire system is under high vacuum.

Hard ionization techniques are processes which impart high quantities of residual energy in the subject molecule invoking large degrees of fragmentation (i.e. the systematic rupturing of bonds acts to remove the excess energy, restoring stability to the resulting ion). Resultant ions tend to have *m/z* lower than the molecular mass (other than in the case of proton transfer and not including isotope peaks). The most common example of hard ionization is electron ionization (EI).

Soft ionization refers to the processes which impart little residual energy onto the subject molecule and as such result in little fragmentation. Examples include fast atom bombardment (FAB), chemical ionization (CI), atmospheric-pressure chemical ionization (APCI), electrospray ionization (ESI), and matrix-assisted laser desorption/ionization (MALDI)

Inductively Coupled Plasma

Inductively coupled plasma ion source.

Inductively coupled plasma (ICP) sources are used primarily for cation analysis of a wide array of sample types. In this source, a plasma that is electrically neutral overall, but that has had a substantial fraction of its atoms ionized by high temperature, is used to atomize introduced sample molecules and to further strip the outer electrons from those atoms. The plasma is usually generated from argon gas, since the first ionization energy of argon atoms is higher than the first of any other elements except He, O, F and Ne, but lower than the second ionization energy of all except the most electropositive metals. The heating is achieved by a radio-frequency current passed through a coil surrounding the plasma.

Other Ionization Techniques

Others include photoionization, glow discharge, field desorption (FD), fast atom bombardment (FAB), thermospray, desorption/ionization on silicon (DIOS), Direct Analysis in Real Time (DART), atmospheric pressure chemical ionization (APCI), secondary ion mass spectrometry (SIMS), spark ionization and al ionization [14]

Mass selection

Mass analyzers separate the ions according to their mass-to-charge ratio. The following two laws govern the dynamics of charged particles in electric and magnetic fields in vacuum.

$\mathbf{F} = Q(\mathbf{E} + \mathbf{v} \times \mathbf{B})$ (Lorentz force law);

F = ***ma*** (Newton's second law of motion in non-relativistic case, i.e. valid only at ion velocity much lower than the speed of light)

Here **F** is the force applied to the ion, *m* is the mass of the ion, **a** is the acceleration, *Q* is the ion charge, **E** is the electric field, and **v** × **B** is the vector cross product of the ion velocity and the magnetic field

Equating the above expressions for the force ap{v} \times \mathbf {B} .}

This differential equation is the classic equation of motion for charged particles. Together with the particle's initial conditions, it completely determines the particle's motion in space and time in terms of *m/Q*. Thus mass spectrometers could be thought of as "mass-to-charge spectrometers". When presenting data, it is common to use the (officially) dimensionless *m/z*, where z is the number of elementary charges (*e*) on the ion (z=Q/e). This quantity, although it is informally called the mass-to-charge ratio, more accurately speaking represents the ratio of the mass number and the charge number, *z*.

There are many types of mass analyzers, using either static or dynamic fields, and magnetic or electric fields, but all operate according to the above differential equation. Each analyzer type has its strengths and weaknesses. Many mass spectrometers use two or more mass analyzers for tandem mass spectrometry (MS/MS). In addition to the more common mass analyzers listed below, there are others designed for special situations.

There are several important analyser characteristics. The mass resolving power is the measure of the ability to distinguish two peaks of slightly different *m/z*. The mass accuracy is the ratio of the *m/z* measurement error to the true m/z. Mass accuracy is usually measured in ppm or milli mass units. The mass range is the range of *m/z* amenable to analysis by a given analyzer. The linear dynamic range is the range over which ion signal is linear with analyte concentration. Speed refers to the time frame of the experiment and ultimately is used to determine the number of spectra per unit time that can be gene

Sector Instruments

ThermoQuest AvantGarde sector mass spectrometer.

A sector field mass analyzer uses an electric and/or magnetic field to affect the path and/or velocity of the charged particles in some way. As shown above, sector instruments bend the trajectories of the ions as they pass through the mass analyzer, according to their mass-to-charge ratios, deflecting the more charged and faster-moving, lighter ions more. The analyzer can be used to select a narrow range of *m/z* or to scan through a range of *m/z* toog the ions pr15]

Time-of-flight

The time-of-flight (TOF) analyzer uses an electric field to accelerate the ions through the same potential, and then measures the time they take to reach the detector. If the particles all have the same charge, the kinetic energies will be identical, and their velocities will depend only on their masses. Ions with a lower mass will the detector first.

Quadrupole Mass Filter

Quadrupole mass analyzers use oscillating electrical fields to selectively stabilize or destabilize the paths of ions passing through a radio frequency (RF) quadrupole field created between 4 parallel rods. Only the ions in a certain range of mass/charge ratio are passed through the system at any time, but changes to the potentials on the rods allow a wide range of m/z values to be swept rapidly, either continuously or in a succession of discrete hops. A quadrupole mass analyzer acts as a mass-selective filter and is closely related to the quadrupole ion trap, particularly the linear quadrupole ion trap except that it is designed to pass the untrapped ions rather than collect the trapped ones, and is for that reason referred to as a transmission quadrupole. A magnetically enhanced quadrupole mass analyzer includes the addition of a magnetic field, either applied axially or transversely. This novel type of instrument leads to an additional performance enhancement in terms of resolution and/or sensitivity depending upon the magnitude and orientation of ted magnetic field.[17][18] A common variation of the transmission quadrupole is the triple quadrupole mass spectrometer. The "triple quad" has three consecutive quadrupole stages, the first acting as a mass filter to transmit a particular incoming ion to the second quadrupole, a collision chamber, wherein that ion can be broken into fragments. The third quadrupole also acts as a mass filter, to transmit a particular fragment ion to the detector. If a quadrupole is made to rapidly and repetitively cycle through a range of mass filter settings, full spectra can be reported. Likewise, a triple quad can be made to perform various scan types characteristic of tandem masrometry.

Ion traps

Three-dimenquadrupole ion trap

The quadrupole ion trap works on the same physical principles as the quadrupole mass analyzer, but the ions are trapped and sequentially ejected. Ions are trapped in a mainly quadrupole RF field, in a space defined by a ring electrode (usually connected to the main RF potential) between two endcap electrodes (typically connected to DC or auxiliary AC potentials). The sample is ionized either internally (e.g. with an electron or laser beam), or externally, in which case the ions are often introduced through an aperture in an endcap electrode.

There are many mass/charge separation and isolation methods but the most commonly used is the mass instability mode in which the RF potential is ramped so that the orbit of ions with a mass $a > b$ are stable while ions with mass b become unstable and are ejected on the z-axis onto a detector. There are also non-destructive analysis methods.

Ions may also be ejected by the resonance excitation method, whereby a supplemental oscillatory excitation voltage is applied to the endcap electrodes, and the trapping voltage amplitude and/or excitation voltage frequency is varied to bring ions into a resonance condition in order oss/charge ratio.

Cylindrical Ion Trap

The cylindrical ion trap mass spectrometer (CIT) is a derivative of the quadrupole ion trap where the electrodes are formed from flat rings rather than hyperbolic shaped electrodes. The architecture lends itself well to miniaturization because as the size of a trap is reduced, the shape of the electric field near the center of the trap, the region where the ions are trapped, forms a shape similar to that of a hyperbolic trap.

Linear Quadrupole ion Trap

A linear quadrupole ion trap is similar to a quadrupole ion trap, but it traps ions in a two dimensional quadrupole field, instead of a three-dimensional quadrupole field as in a 3D quadrupole ion trap. Thermo Fisher`s LTQ ("linear trap quadrupole") is an examp the linear ion trap.

A toroidal ion trap can be visualized as a linear quadrupole curved around and connected at the ends or as a cross section of a 3D ion trap rotated on edge to form the toroid, donut shaped trap. The trap can store large volumes of ions by distributing them throughout the ring-like trap structure. This toroidal shaped trap is a configuration that allows the increased miniaturization of an ion trap mass analyzer. Additionally all ions are stored in the same trapping field and ejected together simplifying detection that can be complicated with array configurations due to variations in detector alignment anhining of the arrays.

As with the toroidal trap, linear traps and 3D quadrupole ion traps are the most commonly miniaturized mass analyzers due to their high sensitivity, tolerance for mTorr pressure, and capabilities for single analyzer tandem mass spectrometry productns).[23]

Orbitrap

Orbitrap mass analyzer

Orbitrap instruments are similar to Fourier transform ion cyclotron resonance mass spectrometers (see text below). Ions are electrostatically trapped in an orbit around a central, spindle shaped electrode. The electrode confines the ions so that they both orbit around the central electrode and oscillate back and forth along the central electrode's long axis. This oscillation generates an image current in the detector plates which is recorded by the instrument. The frequencies of these image currents depend on the mass to charge ratios of the ions. Mass spectra are obtained by Fourier transformation of the recorded image currents.

Orbitraps have a high mass accuracy, high sensitivity good dynamic range.

Fourier Transform ion Cyclotron Resonance

Fourier transform mass spectrometry (FTMS), or more precisely Fourier transform ion cyclotron resonance MS, measures mass by detecting the image current produced by ions cyclotroning in the presence of a magnetic field. Instead of measuring the deflection of ions with a detector such as an electron multiplier, the ions are injected into a Penning trap (a static electric/magnetic ion trap) where they effectively form part of a circuit. Detectors at fixed positions in space measure the electrical signal of ions which pass near them over time, producing a periodic signal. Since the frequency of an ion's cycling is determined by its mass to charge ratio, this can be deconvoluted by performing a Fourier transform on the signal. FTMS has the advantage of high sensitivity (since each ion is "counted" more than once) and much higher resod thus precision.

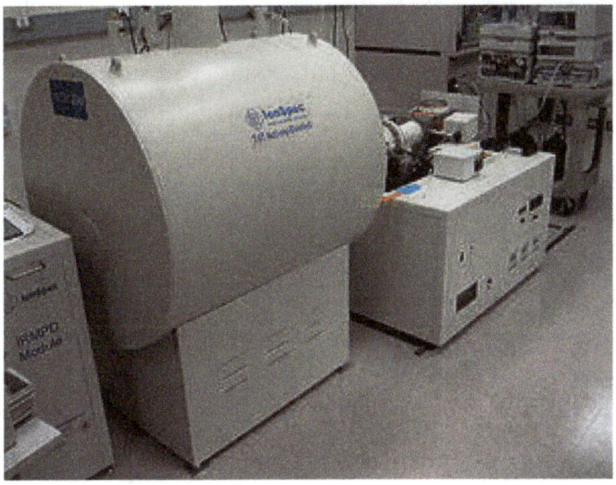

Fourier transform ion cyclotron resonance mass spectrometer

Ion cyclotron resonance (ICR) is an older mass analysis technique similar to FTMS except that ions are detected with a traditional detector. Ions trapped in a Penning trap are excited by an RF electric field until they impact the wall of the trap, where the detector is located. Ions of different mass are resolved according t time.

Detectors

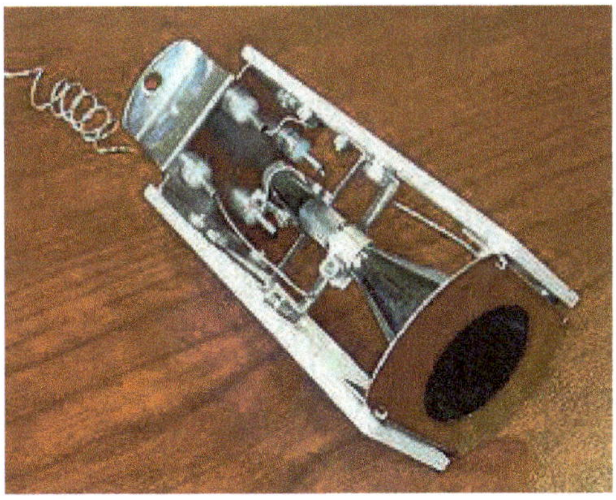

A continuous dynode particle multiplier detector.

The final element of the mass spectrometer is the detector. The detector records either the charge induced or the current produced when an ion passes by or hits a surface. In a scanning instrument, the signal produced in the detector during the course of the scan versus where the instrument is in the scan (at what *m/Q*) will produce a mass spectrum, a record of ions as a function of *m/Q*.

Typically, some type of electron multiplier is used, though other detectors including Faraday cups and ion-to-photon detectors are also used. Because the number of ions leaving the mass analyzer at a particular instant is typically quite small, considerable amplification is often necessary to get a signal. Microchannel plate detectors are commonly used in modeommercial instruments. In FTMS and Orbitraps, the detector consists of a pair of metal surfaces within the mass analyzer/ion trap region which the ions only pass near as they oscillate. No direct current is produced, only a weak AC image current is produced in a circuit between the electrodes. Other inductive detehave also been used.

Tandem Mass Spectrometry

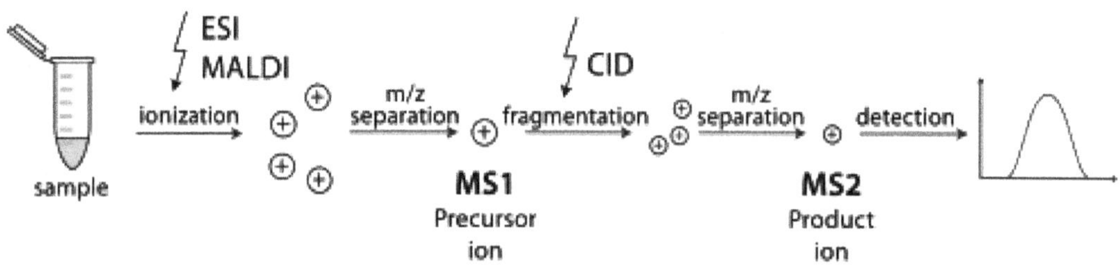

Tandem mass spectrometry for biological molecules using ESI or MALDI.

A tandem mass spectrometer is one capable of multiple rounds of mass spectrometry, usually separated by some form of molecule fragmentation. For example, one mass analyzer can isolate one peptide from many entering a mass spectrometer. A second mass analyzer then stabilizes the peptide ions while they collide with a gas, causing them to fragment by collision-induced dissociation (CID). A third mass analyzer then sorts the fragments produced from the peptides. Tandem MS can also be done in a single mass analyzer over time, as in a quadrupole ion trap. There are various methods for fragmenting molecules for tandem MS, including collision-induced dissociation (CID), electron capture dissociation (ECD), electron transfer dissociation (ETD), infrared multiphoton dissociation (IRMPD), blackbody infrared radiative dissociation (BIRD), electron-detachment dissociation (EDD) and surface-induced dissociation (SID). An important application using tandem mass spectrometry is otein identification.

Tandem mass spectrometry enables a variety of experimental sequences. Many commercial mass spectrometers are designed to expedite the execution of such routine sequences as selected reaction monitoring (SRM) and precursor ion scanning. In SRM, the first analyzer allows only a single mass through and the second analyzer monitors for multiple user-defined fragment ions. SRM is most often used with scanning instruments where the second mass analysis event is duty cycle limited. These experiments are used to increase specificity of detection of known molecules, notably in pharmacokinetic studies. Precursor ion scanning refers to monitoring for a specific loss from the precursor ion. The first and second mass analyzers scan across the spectrum as partitioned by a user-defined *m/z* value. This experiment is used to detect specific motifs within unknown molecules.

Another type of tandem mass spectrometry used for radiocarbon dating is accelerator mass spectrometry (AMS), which uses very high voltages, usually in the mega-volt range, to accelerate negative ions into a type of tandem mass spectrometer.

Common Mass Spectrometer Configurations and Techniques

When a specific configuration of source, analyzer, and detector becomes conventional in practice, often a compound acronym arises to designate it, and the compound acronym may be better known among nonspectrometrists than the component acronyms. The epitome of this is MALDI-TOF, which simply refers to combining a matrix-assisted laser desorption/ionization source with a time-of-flight mass analyzer. The MALDI-TOF moniker is more widely recognized by the non-mass spectrometrists than MALDI or TOF individually. Other examples include inductively coupled plasma-mass spectrometry (ICP-MS), accelerator mass spectrometry (AMS), thermal ionization-mass spectrometry (TIMS) and spark source mass spectrometry (SSMS). Sometimes the use of the generic "MS" actually connotes a very specific mass analyzer and detection system, as is the case with AMS, which is always sector based.

Certain applications of mass spectrometry have developed monikers that although strictly speaking would seem to refer to a broad application, in practice have come instead to connote a specific or a limited number of instrument configurations. An example of this is isotope ratio mass spectrometry (IRMS), which refers in practice to the use of a limited number of sector based mass analyzers; this name is used to refer to both the application and the instrument used for the application.

Separation Techniques Combined with Mass Spectrometry

An important enhancement to the mass resolving and mass determining capabilities of mass spectrometry is using it in tandem with chromatographic and other separation techniques.

Gas Chromatography

Indianapolis Museum of Art conservation scientist performing liquid chromatography–mass spectrometry.

Similar to gas chromatography MS (GC/MS), liquid chromatography-mass spectrometry (LC/MS or LC-MS) separates compounds chromatographically before they are introduced to the ion source and mass spectrometer. It differs from GC/MS in that the mobile phase is liquid, usually a mixture of water and organic solvents, instead of gas. Most commonly, an electrospray ionization source is used in LC/MS. Other popular and commercially available LC/MS ion sources are atmospheric pressure chemical ionization and atmospheric pressure photoionization. There are also some newly developed ionization techniques like laser spray.

Capillary electrophoresis–mass spectrometry

Capillary electrophoresis–mass spectrometry (CE-MS) is a technique that combines the liquid separation process of capillary electrophoresith mass spectrometry.[31] CE-MS is typically coupled ttrospray ion.[32]

Ion Mobility

Ion mobility spectrometry-mass spectrometry (IMS/MS or IMMS) is a technique where ions are first separated by drift time through some neutral gas under an applied electrical potential gradient before being introducedo a mass spectrometer.[33] Drift time is a measure of the radius relative to the charge of the ion. The duty cycle of IMS (the time over which the experiment takes place) is longer than most mass spectrometric techniques, such that the mass spectrometer can sample along the course of the IMS separation. This produces data about the IMS separation and the mass-to-charge ratio of the ions in ner similar to LC/MS.[34]

The duty cycle of IMS is short relative to liquid chromatography or gas chromatography separations and can thus be coupled to such techniques, producing triple mods such as LC/IMS/

Data and Analysis

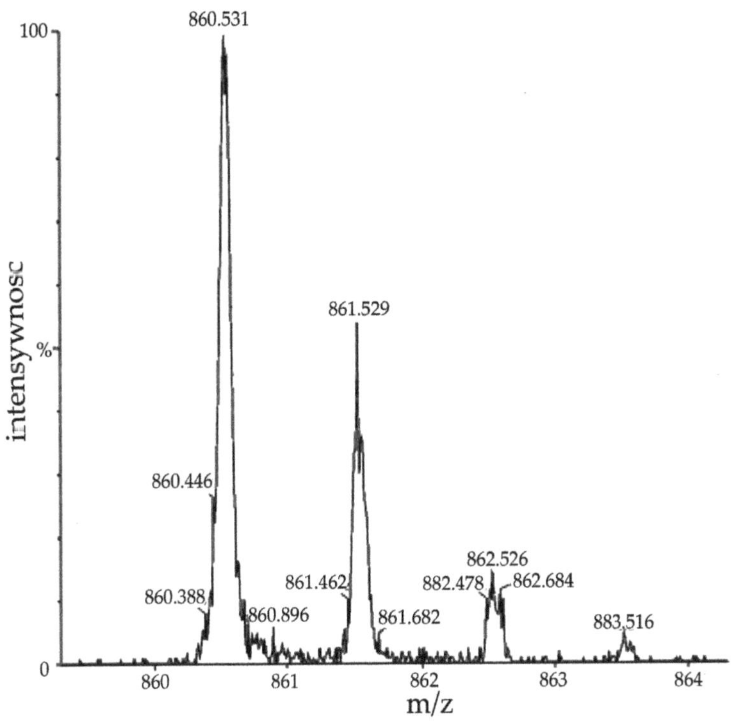

Mass spectrum of a peptide showing the isotopic distribut*ata* ***Representations***

Mass spectrometry produces various types of data. The most common data representation is the mass spectrum.

Certain types of mass spectrometry data are best represented as a mass chromatogram. Types of chromatograms include selected ion monitoring (SIM), total ion current (TIC), and selected reaction monitoring (SRM), among many others.

Other types of mass spectrometry data are well represented as a three-dimensional contour map. In this form, the mass-to-charge, *m/z* is on the *x*-axis, intensity the *y*-axis, and an additional experimental parameter, such as time, is recorded on xis.

Data Analysis

Mass spectrometry data analysis is specific to the type of experiment producing the data. General subdivisions of data are fundamental to understanding any data.

Many mass spectrometers work in either *negative ion mode* or *positive ion mode*. It is very important to know whether the observed ions are negatively or positively charged. This is often important in determining the neutral mass but it also indicates something about the nature of the molecules. Different types of ion source result in different arrays of fragments produced from the original molecules. An electron ionization source produces many fragments and mostly single-charged (1-) radicals (odd number of electrons), whereas an electrospray source usually produces non-radical quasimolecular ions that are frequently multiply charged. Tandem mass spectrometry purposely produces fragment ions post-source and can drastically change the sort of data achieved by an experiment.

Knowledge of the origin of a sample can provide insight into the component molecules of the sample and their fragmentations. A sample from a synthesis/manufacturing process will probably contain impurities chemically related to the target component. A crudely prepared biological sample will probably contain a certain amount of salt, which may form adducts with the analyte molecules in certain analyses.

Results can also depend heavily on sample preparation and how it was run/introduced. An important example is the issue of which matrix is used for MALDI spotting, since much of the energetics of the desorption/ionization event is controlled by the matrix rather than the laser power. Sometimes samples are spiked with sodium or another ion-carrying species to produce adducts rather than a protonated species.

Mass spectrometry can measure molar mass, molecular structure, and sample purity. Each of these questions requires a different experimental procedure; therefore, adequate definition of the experimental goal is a prerequisite for collecting the proper data and successfully interpreting it.

Interpretation of Mass Spectra

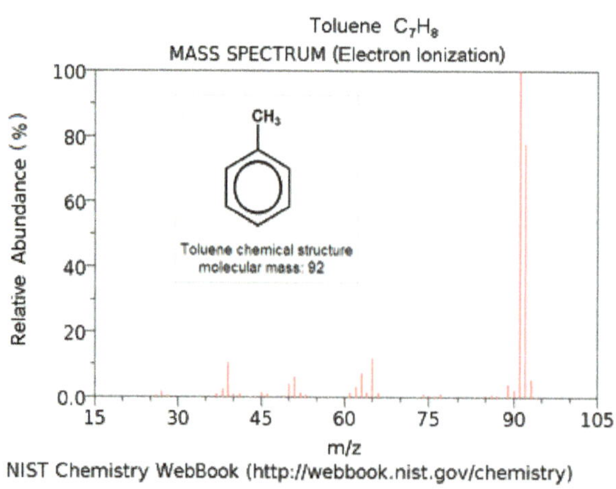

Toluene electron ionization mass spectrum.

Since the precise structure or peptide sequence of a molecule is deciphered through the set of fragment masses, the interpretation of mass spectra requires combined use of various techniques. Usually the first strategy for identifying an unknown compound is to compare its experimental mass spectrum against a library of mass spectra. If no matches result from the search, manual interpretation or software assisted interpretation of mass spectra must be performed. Computer simulation of ionization and fragmentation processes occurring in mass spectrometer is the primary tool for assigning structure or peptide sequence to a molecule. An a priori structural information is fragmented in silico and the resulting pattern is compared with observed spectrum. Such simulation is often supported fragmentation library that contains published patterns of known decomposition reactions. Software taking advantage of this idea has been developed for both small molecules and proteins.

Analysis of mass spectra can also be spectra with accurate mass. A mass-to-charge ratio value (*m/z*) with only integer precision can represent an immense number of theoretically possible ion structures; however, more precise mass figures significantly reduce the number of candidate molecular formulas. A computer algorithm called formula generator calculates all molecular formulas that theoretically fit a given mass with specified tolerance.

A recent technique for structure elucidation in mass spectrometry, called precursor ion fingerprinting, identifies individual pieces of structural information by conducting a search of the tandem spectra of the molecule under investigation against a library of the product-ion spectra of structurally chaized precurs.

Applications

NOAA Particle Analysis by Laser Mass Spectrometry aerosol mass spectrometer aboard a NASA WB-57 high-altitude research aircraft.

Mass spectrometry has both qualitative and quantitative uses. These include identifying unknown compounds, determining the isotopic composition of elements in a molecule, and determining the structure of a compound by observing its fragmentation. Other uses include quantifying the amount of a compound in a sample or studying the fundamentals of gas phase ion chemistry (the chemistry of ions and neutrals in a vacuum). MS is now in very common use in analytical laboratories that study physical, chemical, or biological properties of a great variety of compounds.

As an analytical technique it possesses distinct advantages such as: Increased sensitivity over most other analytical techniques because the analyzer, as a mass-charge filter, reduces background interference, Excellent specificity from characteristic fragmentation patterns to identify unknowns or confirm the presence of suspected compounds, Information about molecular weight, Information about the isotopic abundance of elements, Temporally resolved chemical data.

A few of the disadvantages of the method is that often fails to distinguish between optical and geometrical isomers and the positions of substituent in o-, m- and p- positions in an aromatic ring. Also, its scope is limited in identifying hydrocarbons that produce similar fragmented ions.

Isotope Ratio MS: Isotope Dating and Tracing

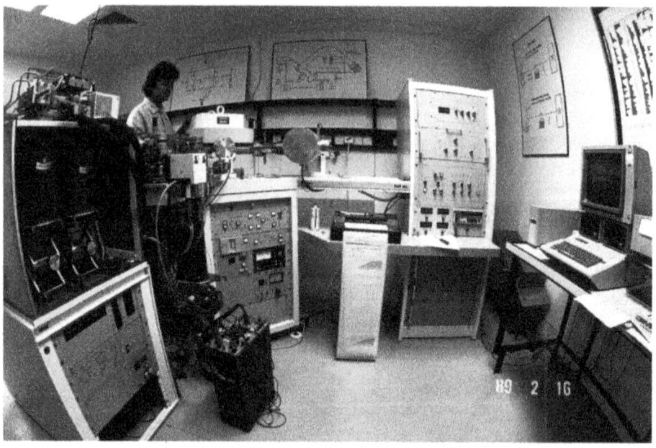

Mass spectrometer to determine the $^{16}O/^{18}O$ and $^{12}C/^{13}C$ isotope ratio on biogenous carbonate.

Mass spectrometry is also used to determine the isotopic composition of elements within a sample. Differences in mass among isotopes of an element are very small, and the less abundant isotopes of an element are typically very rare, so a very sensitive instrument is required. These instruments, sometimes referred to as isotope ratio mass spectrometers (IR-MS), usually use a single magnet to bend a beam of ionized particles towards a series of Faraday cups which convert particle impacts to electric current. A fast on-line analysis of deuterium content of water can be done using flowing afterglow mass spectrometry, FA-MS. Probably the most sensitive and accurate mass spectrometer for this purpose is the accelerator mass spectrometer (AMS). This is because it provides ultimate sensitivity, capable of measuring individual atoms and measuring nuclides with a dynamic range of ~10^{15} relative to major stable isotope.[39]Isotope ratios are important markers of a variety of processes. Some isotope ratios are used to determine the age of materials for example as in carbon dating. Labeling with stable isotopes is also used for protein quantification. (see protein characterization

Trace Gas Analysis

Several techniques use ions created in a dedicated ion source injected into a flow tube or a drift tube: selected ion flow tube (SIFT-MS), and proton transfer reaction (PTR-MS), are variants of chemical ionization dedicated for trace gas analysis of air, breath or liquid headspace using well defined reaction time allowing calculations of analyte concentrations from the known reaction kinetics without the need for internal standard oration.

Atom Probe

An atom probe is an instrument that combines time-of-flight mass spectrometry and field-evaporation microscopy to map the location of individua.

Pharmacokinetics

Pharmacokinetics is often studied using mass spectrometry because of the complex nature of the matrix (often blood or urine) and the need for high sensitivity to observe low dose and long time point data. The most common instrumentation used in this application is LC-MS with a triple quadrupole mass spectrometer. Tandem mass spectrometry is usually employed for added specificity. Standard curves and internal standards are used for quantitation of usually a single pharmaceutical in the samples. The samples represent different time points as a pharmaceutical is administered and then metabolized or cleared from the body. Blank or t=0 samples taken before administration are important in determining background and ensuring data integrity with such complex sample matrices. Much attention is paid to the linearity of the standard curve; however it is not uncommon to use curve fitting with more complex functions such as quadratics since the response of most mass spectrometers is less than linear across lation ranges.

There is currently considerable interest in the use of very high sensitivity mass spectrometry for microdosing studies, which are seen as a promising alternative to animal experimentation.

Protein Characterization

Mass spectrometry is an important method for the characterization and sequencing of proteins. The two primary methods for ionization of whole proteins are electrospray ionization (ESI) and matrix-assisted laser desorption/ionization (MALDI). In keeping with the performance and mass range of available mass spectrometers, two approaches are used for characterizing proteins. In the first, intact proteins are ionized by either of the two techniques described above, and then introduced to a mass analyzer. This approach is referred to as “top-down” strategy of protein analysis. The top-down approach however is largely limited to low-throughput single-protein studies. In the second, proteins are enzymatically digested into smaller peptides using proteases such as trypsin or pepsin, either in solution or in gel after electrophoretic separation. Other proteolytic agents are also used. The collection of peptide products are then introduced to the mass analyzer. When the characteristic pattern of peptides is used for the identification of the protein the method is called peptide mass fingerprinting (PMF), if the identification is performed using the sequence data determined in tandem MS analysis it is called de novo peptide sequencing. These procedures of protein analysis are also referred to as the "bottom-up" approach. A third approach however is beginning to be used, this intermediate "middle-down" approach involves analyzing proteolytic peptide larger than tical tryptic peptide.

Glycan Analysis

Mass spectrometry (MS), with its low sample requirement and high sensitivity, has been predominantly used in glycobiology for characterization and elucidn of glycan structures. Mass spectrometry provides a complementary method to HPLC for the analysis of glycans. Intact glycans may be detected directly as singly charged ions by matrix-assisted laser desorption/ionization mass spectrometry (MALDI-MS) or, following permethylation or peracetylation, by fast atom bombardment spectrometry (FAB-MS). Electrospray ionization mass spectrometry (ESI-MS) also gives good signaor the smaller glycans. Various free and commercial software are now available which interpret MS data and aid in Glycan structure character.

Space Exploration

NASA's Phoenix Mars Lander analyzing a soil sample from the "Rosy Red" trench with the TEGA mass spectrometer.

As a standard method for analysis, mass spectrometers have reached other planets and moons. Two were taken to Mars by the Viking program. In early 2005 the Cassini–Huygens mission delivered a specialized GC-MS instrument aboard the Huygens probe through the atmosphere of Titan, the largest moon of the planet Saturn. This instrument analyzed atmospheric samples along its descent trajectory and was able to vaporize and analyze samples of Titan's frozen, hydrocarbon covered surface once the probe had landed. These measurements compare the abundance of isotope(s) of each particle comparatively toth's natural abundance. Also on board the Cassini–Huygens spacecraft is an ion and neutral mass spectrometer which has been taking measurements of Titan's atmospheric composition as well as the composition of Enceladus' plumes. A Thermal and Evolved Gas Analyzer mass spectrometer was carried by the Mars Phoennder launched in 2007.

Mass spectrometers are also widely used in space missions to measure the composition of plasmas. For example, the Cassini spacecraft carries the Cassini ma Spectrometer (CAPS), which measures the mass of ions in Saturn's magnetosphere.

Respired Gas Monitor

Mass spectrometers were used in hospitals for respiratory gas analysis beginning around 1975 through the end of the century. Some are probably still in use but none are culy being manufactured.

Found mostly in the operating room, they were a part of a complex system, in which respired gas samples from patients undergoing anesthesia were drawn into the instrument through a valve mechanism designed to sequentially connect up to 32 rooms to the mass spectrometer. A computer directed all operations of the system. The data collected from the mass spectrometer was delivered to the individual rooms for the anesthesiologist to use.

The uniqueness of this magnetic sector mass spectrometer may have been the fact that a plane of detectors, each purposely positioned to collect all of the ion species expected to be in the samples, allowed the instrument to simultaneously report all of the gases respired by the patient. Although the mass range was limited to slightly over 120 u, fragmentation of some of the heavier molecules negated the need fogher detection limit.

Preparative Mass Spectrometry

The primary function of mass spectrometry is as a tool for chemical analyses based on detection and quantification of ions according to their mass-to-charge ratio. However, mass spectrometry also shows promfor material synthesis. Ion soft landing is characterized by deposition of intact species on surfaces at low kinetic energies which precludes the fragmentatif the incident species. The soft landing technique was first reported in 1977 for the reaction of low energy sulfur containns on a lead surface.

REFERENCES

1. Sparkman, O. David (2000). Mass spectrometry desk reference. Pittsburgh: Global View Pub. ISBN 0-9660813-2-3.

2. "Definition of spectrograph." Merriam Webster. Accessed 13 June 2008.
3. Downard, Kevin (2004).Mass Spectrometry - A Foundation Course". Royal Society of Chemistry. doi:10.1039/9781847551306. ISBN 978-0-85404-609-6.
4. Squires, Gordon (1998).Francis Aston and the mass spectrograph". Dalton Transactions *(23): 3893–3900.* doi:10.1039/a804629h.
5. KM Downard (2007).Francis William Aston – the man behind the mass spectrograph". European Journal of Mass Spectrometry. **13** *(3): 177–190.* doi:10.1255/ejms.878. PMID 17881785.
6. Thomson, J.J. *(1913).* Rays Of Positive Electricity and Their Application to Chemical Analysis. London: Longman's Green and Company.
7. Siri, William (1947).Mass spectroscope for analysis in the low-mass range". Review of Scientific Instruments. **18** *(8): 540–545.* Bibcode:1947RScI...18..540S. doi:10.1063/1.1740998.
8. Price, Phil (1991).Standard definitions of terms relating to mass spectrometry. A report from the Committee on Measurements and Standards of the American Society for Mass Spectrometry". Journal of the American Society for Mass Spectrometry. **2** *(4): 336–348.* doi:10.1016/1044-0305(91)80025-3. PMID 24242353.
9. Parkins, William E. (2005).The uranium bomb, the calutron, and the space-charge problem". Physics Today. **58** *(5): 45–51.* Bibcode:2005PhT....58e..45P. doi:10.1063/1.1995747. ISSN 0031-9228.
10. The Nobel Prize in Chemistry 2002: Information for the Public". The Nobel Foundation. 9 October 2002. Retrieved 2007-08-29.
11. Fenn, J. B.; Mann, M.; Meng, C. K.; Wong, S. F.; Whitehouse, C. M. (1989). Electrospray ionization for mass spectrometry of large biomolecules". Science. **246** *(4926): 64–71.* Bibcode:1989Sci...246...64F. doi:10.1126/science.2675315. PMID 2675315.
12. Tanaka, K.; Waki, H.; Ido, Y.; Akita, S.; Yoshida, Y.; Yoshida, T. (1988). Protein and Polymer Analyses up to m/z 100 000 by Laser Ionization Time-of flight Mass Spectrometry". Rapid Commun Mass Spectrom. **2** *(20): 151–3.* doi:10.1002/rcm.1290020802.
13. Karas, M.; Bachman, D.; Bahr, U.; Hillenkamp, F. (1987).Matrix-Assisted Ultraviolet Laser Desorption of Non-Volatile Compounds". Int J Mass Spectrom Ion Proc. **78**: 53–68. Bibcode:1987IJMSI..78...53K. doi:10.1016/0168-1176(87)87041-6.
14. Bruins, A. P. (1991).Mass spectrometry with ion sources operating at atmospheric pressure". Mass Spectrometry Reviews. **10** *(1): 53–77.* doi:10.1002/mas.1280100104.
15. Cottrell, John S & Greathead, Roger J (1986).Extending the Mass Range of a Sector Mass Spectrometer". Mass Spectrometry Reviews. **5** *(3): 215–247.* doi:10.1002/mas.1280050302.

16. In the event that the ions do not start at identical kinetic energies, some ions may lag behind higher kinetic energy ions decreasing resolution. Reflectron geometries are commonly employed to correct this problem. Wollnik, H. (1993). Time-of-flight mass analyzers". Mass Spectrometry Reviews. **12** *(2): 89–114.* doi:10.1002/mas.1280120202.

17. Syed S; Maher S; Taylor S (2013).Quadrupole mass filter operation under the influence of magnetic field". Journal of Mass Spectrometry. **48** *(12): 1325–1339.* doi:10.1002/jms.3293.

18. Maher S; Syed S; Hughes D; Gibson J; Taylor S (2013).Mapping the stability diagram of a quadrupole mass spectrometer with a static transverse magnetic field applied". Journal of the American Society for Mass Spectrometry. **24** *(8): 1307–1314.* Bibcode:2013JASMS..24.1307M. doi:10.1007/s13361-013-0654-5.

19. Paul, W.; Steinwedel, H. (1953).Ein neues Massenspektrometer ohne Magnetfeld". Zeitschrift für Naturforschung A. **8** *(7): 448–450.* Bibcode:1953ZNatA...8..448P. doi:10.1515/zna-1953-0710.

20. March, R. E. (2000).Quadrupole ion trap mass spectrometry: a view at the turn of the century". International Journal of Mass Spectrometry. **200** *(1–3): 285–312.* Bibcode:2000IJMSp.200..285M. doi:10.1016/S1387-3806(00)00345-6.

21. Schwartz, Jae C.; Senko, Michael W.; Syka, John E. P. (2002).A two-dimensional quadrupole ion trap mass spectrometer". Journal of the American Society for Mass Spectrometry. **13** *(6): 659–669.* doi:10.1016/S1044-0305(02)00384-7. PMID 12056566.

22. Lammert SA, Rockwood AA, Wang M, Lee ML, Lee ED, Tolley SE, Oliphant JR, Jones JL, Waite RW (2006).Miniature Toroidal Radio Frequency Ion Trap Mass Analyzer". Journal of the American Society for Mass Spectrometry. **17** *(7): 916–922.* doi:10.1016/j.jasms.2006.02.009. PMID 16697659.

23. Snyder, Dalton; Pulliam, Christopher; Ouyang, Zheng; Cooks, R. Graham (2016).Miniature and fieldable mass spectrometers: Recent advances". Anal. Chem. **88**: 2–29. doi:10.1021/acs.analchem.5b03070.

24. Hu, Qizhi; Noll, Robert J.; Li, Hongyan; Makarov, Alexander; Hardman, Mark; Graham Cooks, R. (2005).The Orbitrap: a new mass spectrometer". Journal of Mass Spectrometry. **40** *(4): 430–443.* doi:10.1002/jms.856. PMID 15838939.

25. Comisarow, M. B. & Marshall, A. G. (1974).Fourier transform ion cyclotron resonance spectroscopy". Chemical Physics Letters. **25** *(2): 282–283.* Bibcode:1974CPL....25..282C. doi:10.1016/0009-2614(74)89137-2.

26. Marshall, A. G.; Hendrickson, C. L.; Jackson, G. S. (1998).Fourier transform ion cyclotron resonance mass spectrometry: a primer". Mass Spectrometry Reviews. **17** *(1): 1–34.* doi:10.1002/(SICI)1098-2787(1998)17:1<1::AID-MAS1>3.0.CO;2-K. PMID 9768511.

27. Dubois, F.; Knochenmuss, R.; Zenobi, R.; Brunelle, A.; Deprun, C.; Le Beyec, Y. (1999).A comparison between ion-to-photon and microchannel plate detectors". Rapid Communications in Mass Spectrometry. **13** *(9): 786–791.* doi:10.1002/(SICI)1097-0231(19990515)13:9<786::AID-RCM566>3.0.CO;2-3.

28. Park, Melvin A.; Callahan, John H.; Vertes, Akos (1994).An inductive detector for time-of-flight mass spectrometry". Rapid Communications in Mass Spectrometry. **8** *(4): 317–322.* doi:10.1002/rcm.1290080407.

29. Boyd, Robert K. (1994).Linked-scan techniques for MS/MS using tandem-in-space instruments". Mass Spectrometry Reviews. **13** *(5–6): 359–410.* doi:10.1002/mas.1280130502.

30. Eiceman, G.A. (2000). Gas Chromatography. In R.A. Meyers (Ed.), Encyclopedia of Analytical Chemistry: Applications, Theory, and Instrumentation, pp. 10627. Chichester: Wiley. ISBN 0-471-97670-9

31. Loo JA, Udseth HR, Smith RD (June 1989).Peptide and protein analysis by electrospray ionization-mass spectrometry and capillary electrophoresis-mass spectrometry". Anal. Biochem. **179** *(2): 404–12.* doi:10.1016/0003-2697(89)90153-X. PMID 2774189.

32. Maxwell EJ, Chen DD (October 2008). "Twenty years of interface development for capillary electrophoresis-electrospray ionization-mass spectrometry". Anal. Chim. Acta. **627** *(1): 25–33.* doi:10.1016/j.aca.2008.06.034. PMID 18790125.

33. Verbeck GF, Ruotolo BT, Sawyer HA, Gillig KJ, Russell DH (2002). "A fundamental introduction to ion mobility mass spectrometry applied to the analysis of biomolecules". J Biomol Tech. **13** *(2): 56–61.* PMC 2279851. PMID 19498967.

34. Matz, Laura M.; Asbury, G. Reid; Hill, Herbert H. (2002).Two-dimensional separations with electrospray ionization ambient pressure high-resolution ion mobility spectrometry/quadrupole mass spectrometry". Rapid Communications in Mass Spectrometry. **16** *(7): 670–675.* doi:10.1002/rcm.623. PMID 11921245.

35. Sowell, Renã A.; Koeniger, Stormy L.; Valentine, Stephen J.; Moon, Myeong Hee; Clemmer, David E. (2004).Nanoflow LC/IMS-MS and LC/IMS-CID/MS of Protein Mixtures". Journal of the American Society for Mass Spectrometry. **15** *(9): 1341–1353.* doi:10.1016/j.jasms.2004.06.014. PMID 15337515.

36. Tureček, František; McLafferty, Fred W. (1993). Interpretation of mass spectra. Sausalito: University Science Books. ISBN 0-935702-25-3.

37. Mistrik, R. (2004). A New Concept for the Interpretation of Mass Spectra Based on a Combination of a Fragmentation Mechanism Database and a Computer Expert System. in Ashcroft, A.E., Brenton, G., Monaghan,J.J. (Eds.), Advances in Mass Spectrometry, Elsevier, Amsterdam, vol. 16, pp. 821.[dead link]

38. Sheldon, Michelle T.; Mistrik, Robert; Croley, Timothy R. (2009).Determination of ion structures in structurally related compounds using precursor ion fingerprinting". Journal of the American Society for Mass Spectrometry. **20** *(3): 370–376.* doi:10.1016/j.jasms.2008.10.017. PMID 19041260.

39. ^ Jump up to:a b Maher S, Jjunju FP, Taylor S (2015).100 years of mass spectrometry: Perspectives and future trends". Rev. Mod. Phys. **87** *(1): 113–135.* Bibcode:2015RvMP...87..113M. doi:10.1103/RevModPhys.87.113.

40. Hsieh, Yunsheng; Korfmacher, WA (2006).Systems for Drug Metabolism and Pharmacokinetic Screening, Y. Hsieh and W.A. Korfmacher,

Current Drug Metabolism". Current Drug Metabolism. **7** *(5): 479–489.* doi:10.2174/138920006777697963. PMID 16787157.

41. Covey, T.R.; Lee, E.D.; Henion, J.D. (1986).Mass Spectrometry for the Determination of Drugs in Biological Samples". Anal. Chem. **58** *(12): 2453–2460.* doi:10.1021/ac00125a022. PMID 3789400.

42. Covey, Tom R.; Crowther, Jonathan B.; Dewey, Elizabeth A.; Henion, Jack D. (1985).Mass Spectrometry Determination of Drugs and Their Metabolites in Biological Fluids". Anal. Chem. **57** *(2): 474–81.* doi:10.1021/ac50001a036. PMID 3977076.

43. Chait, Brian T. (2011). "Mass Spectrometry in the Postgenomic Era". **80**: 239–46. doi:10.1146/annurev-biochem-110810-095744. PMID 21675917. – via Annual Reviews (subscription required)

44. Apte A, Meitei NS (2009).Bioinformatics in Glycomics: Glycan Characterization with Mass Spectrometric Data Using SimGlycan". Methods in molecular biology. Methods in Molecular Biology. **600**: 269–281. doi:10.1007/978-1-60761-454-8_19. ISBN 978-1-60761-453-1. PMID 19882135.

45. Harvey, D.; Dwek, R.A.; Rudd, P.M. (2000).Determining the Structure of Glycan Moieties by Mass Spectrometry". Current Protocols *in Protein Science. Chapter 12: 12.7 12.7.15.* doi:10.1002/0471140864.ps1207s43. ISBN 0-471-14086-4. PMID 18429296.

46. Blow, Nathan (2009).Glycobiology: A spoonful of sugar". Nature. **457** *(7229): 617–620.* Bibcode:2009Natur.457..617B. doi:10.1038/457617a. PMID 19177129.

47. Petrie, S. & Bohme, D. K. (2007).Ions in space". Mass Spectrometry Reviews. **26** *(2): 258–280.* doi:10.1002/mas.20114. PMID 17111346.

48. Hoffman, J; Chaney, R; Hammack, H (2008).Phoenix Mars Mission—The Thermal Evolved Gas Analyzer". Journal of the American Society for Mass Spectrometry. **19** *(10): 1377–83.* doi:10.1016/j.jasms.2008.07.015. PMID 18715800.

49. Cassini Plasma Spectrometer". Southwest Research Institute. Retrieved 2008-01-04.

50. Riker JB, Haberman B (1976).Expired gas monitoring by mass spectrometry in a respiratory intensive care unit". Crit. Care Med. **4** *(5): 223–9.* doi:10.1097/00003246-197609000-00002. PMID 975846.

51. Gothard, J.W.W.; Busst, C.M.; Branthwaite, M.A.; Davies, N.J.H.; Denison, D.M. (1980).Applications of respiratory mass spectrometry to intensive care". Anaesthesia. **35** *(9): 890–895.* doi:10.1111/j.1365-2044.1980.tb03950.x. PMID 6778243.

52. Verbeck G, Hoffmann W, Walton B (2012).Soft-landing preparative mass spectrometry". Analyst. **137** *(19): 4393–4407.* Bibcode:2012Ana...137.4393V. doi:10.1039/C2AN35550G.

53. Franchetti V, Solka BH, Baitinger WE, Amy JW, Cooks RG (1977).Soft landing of ions as a means of surface modification". Mass Spectrom. Ion Phys. **23** *(1): 29–35.* Bibcode:1977IJMSI..23...29F. doi:10.1016/0020-7381(77)80004-1

CHAPTER 9

Electrochromatography

Chapter Contents

- **Introduction**
- **Micellar Electrokinetic Chromatography**
- **Theory**
- **History**
- **Applications of Mekc**
- **Displacement Chromatography**

Electrochromatography is a separation method of mixture of chemical compounds which uses simultanously occurring interphase reactions and electric field influence. Theelectroosmotically driven flow is employed in the discussed systems. Sometimes a pressure-driven flow is also used. Various types of interactions, mentioned above provide the new separation possibilities. However, a selection of separation conditions cannot be only limited to generally known methods as chromatography and electrophoresis.

INTRODUCTION

Electrochromatography is a chemical separation technique in analytical chemistry, biochemistry and molecular biology used to resolve and separate mostly large biomolecules such as proteins. It is a combination of size exclusion chromatography (gel filtration chromatography) and gel electrophoresis. These separation mechanisms operate essentially in superposition along the length of a gel filtration column to which an axial electric field gradient has been added. The molecules are separated by size due to the gel filtration mechanism and by electrophoretic mobility due to the gel electrophoresis mechanism. Additionally there are secondary chromatographic solute retention mechanisms.

MICELLAR ELECTROKINETIC CHROMATOGRAPHY

Micellar electrokinetic chromatography (MEKC), is a chromatography technique, used in analytical chemistry. It is a modification of capillary electrophoresis (CE), where the samples are separated by differential partitioning between micelles (pseudo-stationary phase) and a surrounding aqueous buffer solution (mobile phase).

The basic set-up and detection methods used for MEKC are the same as those used in CE. The difference is that the solution contains a surfactant at a concentration that is greater than the critical micelle concentration (CMC). Above this concentration, surfactant monomers are in equilibrium with micelles.

In most applications, MEKC is performed in open capillaries under alkaline conditions to generate a strong electroosmotic flow. Sodium dodecyl sulfate (SDS) is the most commonly used surfactant in MEKC applications. The anionic character of the sulfate groups of SDS cause the surfactant and micelles to have electrophoretic mobility that is counter to the direction of the strong electroosmotic flow. As a result, the surfactant monomers and micelles migrate quite slowly, though their net movement is still toward the cathode. During a MEKC separation, analytes distribute themselves between the hydrophobic interior of the micelle and hydrophilic buffer solution as shown in *figure 1*.

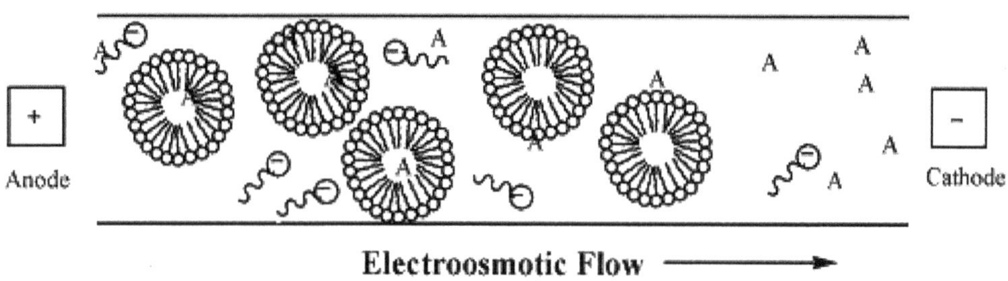

Figure 1: Depiction of the distribution of analytes (A) based on their hydrophobicity.

Analytes that are insoluble in the interior of micelles should migrate at the electroosmotic flow velocity, u_o, and be detected at the retention time of the buffer, t_M. Analytes that solubilize completely within the micelles (analytes that are highly hydrophobic) should migrate at the micelle velocity, u_c, and elute at the final elution time, t_c.

THEORY

The micelle velocity is defined by:

$$u_c = u_p + u_o$$

where u_p is the electrophoretic velocity of a micelle.

The retention time of a given sample should depend on the capacity factor, n_w:

$$k^1 = \frac{n_c}{n_w}$$

where n_c is the total number of moles of solute in the micelle and n_w is the total moles in the aqueous phase. The retention time of a solute should then be within the range:

$$t_M \leq t_r \leq t_c$$

Charged analytes have a more complex interaction in the capillary because they exhibit electrophoretic mobility, engage in electrostatic interactions with the micelle, and participate in hydrophobic partitioning.

The fraction of the sample in the aqueous phase, R, is given by:

$$R = \frac{u_s - u_c}{u_o - u_c}$$

where u_s is the migration velocity of the solute. The value R can also be expressed in terms of the capacity factor:

$$R = \frac{1}{1 + k^1}$$

Using the relationship between velocity, tube length from the injection end to the detector cell L, and retention time, $u_o = L/t_M$, $u_c = L/t_c$ and $u_s = L/t_r$, a relationship between the capacity factor and retention times can be formulated:

$$k^1 = \frac{t_r - t_M}{t_M(1 - (t_r/t_c))}$$

The extra term enclosed in parenthesis accounts for the partial mobility of the hydrophobic phase in MEKC. This equation resembles an expression derived for k^1 in conventional packed bed chromatography:

$$k = \frac{t_r - t_M}{t_M}$$

A rearrangement of the previous equation can be used to write an expression for the retention factor:

$$t_r = \left(\frac{1 + k^1}{1 + (t_M/t_c)k^1}\right) t_M$$

From this equation it can be seen that all analytes that partition strongly into the micellar phase (where k^1 is essentially ∞) migrate at the same time, t_c. In conventional chromatography, separation of similar compounds can be improved by gradient elution. In MEKC, however, techniques must be used to extend the elution range to separate strongly retained analytes.

Elution ranges can be extended by several techniques including the use of organic modifiers, cyclodextrins, and mixed micelle systems. Short-chain alcohols or acetonitrile can be used as organic modifiers that decrease t_M and k^1 to improve the resolution of analytes that co-elute with the micellar phase. These agents, however, may alter the level of the EOF. Cyclodextrins are cyclic polysaccharides that form inclusion complexes that can cause competitive hydrophobic partitioning of the analyte. Since analyte-cyclodextrin complexes are neutral, they will migrate toward the cathode at a higher velocity than that of the negatively charged micelles. Mixed micelle systems, such as the one formed by combining SDS with the neutral surfactant Brij-35, can also be used to alter the selectivity of MEKC.

HISTORY

In 1984, the Terabe group reported a technique that enabled capillary electrophoresis (CE) instrumentation to be used in the separation of neutral (as well as ionic) species.

APPLICATIONS OF MEKC

The simplicity and efficiency of MEKC have made it an attractive technique for a variety of applications. Further improvements can be made to the selectivity of MEKC by adding chiral selectors or chiral surfactants to the system. Unfortunately, this technique is not suitable for protein analysis because proteins are generally too large to partition into a surfactant micelle and tend to bind to surfactant monomers to form SDS-protein complexes.

Recent applications of MEKC include the analysis of uncharged pesticides, essential and branched-chain amino acids in nutraceutical products, and hydrocarbon and alcohol contents of the marjoram herb.

MEKC has also been targeted for its potential to be used in combinatorial chemical analysis. The advent of combinatorial chemistry has enabled medicinal chemists to synthesize and identify large numbers of potential drugs in relatively short periods of time. Small sample and solvent requirements and the high resolving power of MEKC have enabled this technique to be used to quickly analyze a large number of compounds with good resolution.

Traditional methods of analysis, like high-performance liquid chromatography (HPLC), can be used to identify the purity of a combinatorial library, but assays need to be rapid with good resolution for all components to provide useful information for the chemist. The introduction of surfactant to traditional capillary electrophoresis instrumentation has dramatically expanded the scope of analytes that can be separated by capillary electrophoresis.

MEKC can also be used in routine quality control of antibiotics in pharmaceuticals or feedstuffs.

DISPLACEMENT CHROMATOGRAPHY

Displacement chromatography is a chromatography technique in which a sample is placed onto the head of the column[n 1] and is then displaced by a solute that is more strongly sorbed than the components of the original mixture. The result is that the components are resolved into consecutive "rectangular" zones of highly concentrated pure substances rather than solvent-separated "peaks". It is primarily a preparative technique; higher product concentration, higher purity, and increased throughput may be obtained compared to other modes of chromatography.

Discovery

The advent of displacement chromatography can be attributed to Arne Tiselius, who in 1943 first classified the modes of chromatography as frontal, elution, and displacement. Displacement chromatography found a variety of applications including isolation of transuranic elements and biochemical entities. The technique was redeveloped by Csaba Horváth, who employed modern high-pressure columns and equipment. It has since found many applications, particularly in the realm of biological macromolecule purification.

Principle

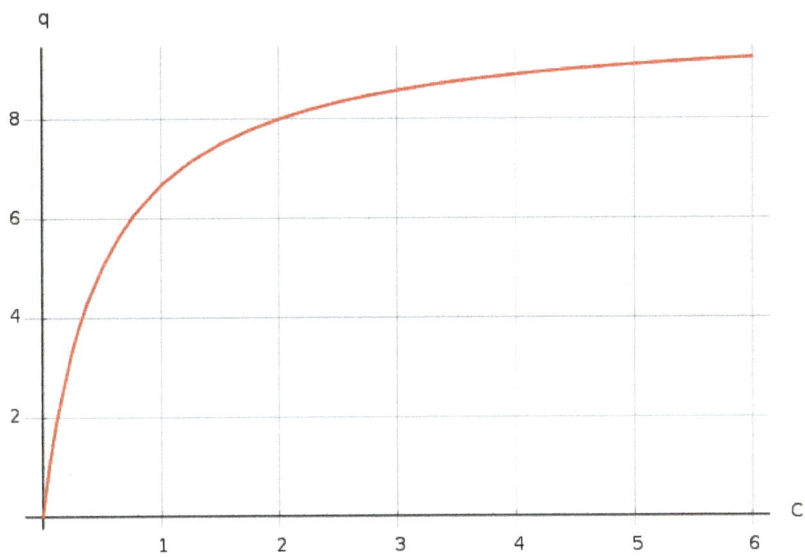

Example of a Langmuir isotherm, for a population of binding sites having uniform affinity. In this case the vertical axis represents the amount bound per unit of stationary phase, the horizontal axis the concentration in the mobile phase. In this case, the dissociation constant is 0.5 and the capacity 10; units are arbitrary.

The basic principle of displacement chromatography is: there are only a finite number of binding sites for solutes on the matrix (the stationary phase), and if a site is occupied by one molecule, it is unavailable to others. As in any chromatography, equilibrium is established between molecules of a given kind bound to the matrix and those of the same kind free in solution. Because the number of binding sites is finite, when the concentration of molecules free in solution is large relative to the dissociation constant for the sites, those sites will mostly be filled. This results in a downward-curvature in the plot of bound *vs* free solute, in the simplest case giving a Langmuir isotherm[n 2]. A molecule with a high affinity for the matrix (the displacer) will compete more effectively for binding sites, leaving the mobile phase enriched in the lower-affinity solute. Flow of mobile phase through the column preferentially carries off the lower-affinity solute and thus at high concentration the higher-affinity solute will eventually displace all molecules with lesser affinities.

Mode of Operation

Loading

At the beginning of the run, a mixture of solutes to be separated is applied to the column, under conditions selected to promote high retention. The higher-affinity solutes are preferentially retained near the head of the column, with the lower-affinity solutes moving farther downstream. The fastest moving component begins to form a pure zone downstream. The other components also begin to form zones, but the continued supply of the mixed feed at head of the column prevents full resolution.

Displacement

After the entire sample is loaded, the feed is switched to the displacer, chosen to have higher affinity than any sample component.[n 4] The displacer forms a sharp-edged zone at the head of the column, pushing the other components downstream. Each sample component now acts as a displacer for the lower-affinity solutes, and the solutes sort themselves out into a series of contiguous bands (a “displacement train”), all moving downstream at the rate set by the displacer. The size and loading of the column are chosen to let this sorting process reach completion before the components reach the bottom of the column. The solutes appear at the bottom of the column as a series of contiguous zones, each consisting of one purified component, with the concentration within each individual zone effectively uniform.

Regeneration

After the last solute has been eluted, it is necessary to strip the displacer from the column. Since the displacer was chosen for high affinity, this can pose a challenge. On reverse-phase materials, a wash with a high percentage of organic solvent may suffice. Large pH shifts are also often employed. One effective strategy is to remove the displacer by chemical reaction; for instance if hydrogen ion was used as displacer it can be removed by reaction with hydroxide, or a polyvalent metal ion can be removed by reaction with a chelating agent. For some matrices, reactive groups on the stationary phase can be titrated to temporarily eliminate the binding sites, for instance weak-acid ion exchangers or chelating resins can be converted to the protonated form. For gel-type ion exchangers, selectivity reversal at very high ionic strength can also provide a solution. Sometimes the displacer is specifically designed with a titratable functional group to shift its affinity. After the displacer is washed out, the column is washed as needed to restore it to its initial state for the next run.

Comparison with Elution Chromatography

Common Fundamentals

In any form of chromatography, the rate at which the solute moves down the column is a direct reflection of the percentage of time the solute spends in the mobile phase. To achieve separation in either elution or displacement chromatography, there must be appreciable differences in the affinity of the respective solutes for the stationary phase. Both methods rely on movement down the column to amplify the effect of small differences in distribution between the two phases. Distribution between the mobile and stationary phases is described by the binding isotherm, a plot of solute bound to (or partitioned into) the stationary phase as a function of concentration in the mobile phase. The isotherm is often linear, or approximately so, at low concentrations, but commonly curves (concave-downward) at higher concentrations as the stationary phase becomes saturated.

Characteristics of Elution Mode

In elution mode, solutes are applied to the column as narrow bands and, at low concentration, move down the column as approximately Gaussian peaks. These peaks continue to broaden as they travel, in proportion to the square root of the distance traveled. For two substances to be resolved, they must migrate down the column at sufficiently different rates to overcome the effects of band spreading. Operating at high concentration, where the isotherm is curved, is disadvantageous in elution chromatography because the rate of travel then depends on concentration, causing the peaks to spread and distort.

Retention in elution chromatography is usually controlled by adjusting the composition of the mobile phase (in terms of solvent composition, pH, ionic strength, and so forth) according to the type of stationary phase employed and the particular solutes to be separated. The mobile phase components generally have lower affinity for the stationary phase than do the solutes being separated, but are present at higher concentration and achieve their effects due to mass action. Resolution in elution chromatography is generally better when peaks are strongly retained, but conditions that give good resolution of early peaks lead to long run-times and excessive broadening of later peaks unless gradient elution is employed. Gradient equipment adds complexity and expense, particularly at large scale.

Advantages and Disadvantages of Displacement Mode

In contrast to elution chromatography, solutes separated in displacement mode form sharp-edged zones rather than spreading peaks. Zone boundaries in displacement chromatography are self-sharpening: if a molecule for some reason gets ahead of its band, it enters a zone in which it is more strongly retained, and will then run more slowly until its zone catches up. Furthermore, because displacement chromatography takes advantage of the non-linearity of the isotherms, loadings are deliberately high; more material can be separated on a given column, in a given time, with the purified components recovered at significantly higher concentrations. Retention conditions can still be adjusted, but the displacer controls the migration rate of the solutes. The displacer is selected to have higher affinity for the stationary phase than does any of the solutes being separated, and its concentration is set to approach saturation of the stationary phase and to give the desired migration rate of the concentration wave. High-retention conditions can be employed without gradient operation, because the displacer ensures removal of all solutes of interest in the designed run time.

Because of the concentrating effect of loading the column under high-retention conditions, displacement chromatography is well suited to purify components from dilute feed streams. However, it is also possible to concentrate material from a dilute stream at the head of a chromatographic column and then switch conditions to elute the adsorbed material in conventional isocratic or gradient modes. Therefore, this approach is not unique to displacement chromatography, although the higher loading capacity and less dilution allow greater concentration in displacement mode.

A disadvantage of displacement chromatography is that non-idealities always give rise to an overlap zone between each pair of components; this mixed zone must be collected separately for recycle or discard to preserve the purity of the separated materials. The strategy of adding spacer molecules to form zones between the components (sometimes termed "carrier displacement chromatography") has been investigated and can be useful when suitable, readily removable spacers are found. Another disadvantage is that the raw chromatogram, for instance a plot of absorbance or refractive index *vs* elution volume, can be difficult to interpret for contiguous zones, especially if the displacement train is not fully developed. Documentation and troubleshooting may require additional chemical analysis to establish the distribution of a given component. Another disadvantage is that the time required for regeneration limits throughput.

According to John C. Ford`s article in the *Encyclopedia of Chromatography*, theoretical studies indicate that at least for some systems, optimized overloaded elution chromatography offers higher throughput than displacement chromatography, though limited experimental tests suggest that displacement chromatography is superior (at least before consideration of regeneration time).

Applications

Historically, displacement chromatography was applied to preparative separations of amino acids and rare earth elements and has also been investigated for isotope separation.

Proteins

The chromatographic purification of proteins from complex mixtures can be quite challenging, particularly when the mixtures contain similarly retained proteins or when it is desired to enrich trace components in the feed. Further, column loading is often limited when high resolutions are required using traditional modes of chromatography (e.g. linear gradient, isocratic chromatography). In these cases, displacement chromatography is an efficient technique for the purification of proteins from complex mixtures at high column loadings in a variety of applications.

An important advance in the state of the art of displacement chromatography was the development of low molecular mass displacers for protein purification in ion exchange systems. This research was significant in that it represented a major departure from the conventional wisdom that large polyelectrolyte polymers are required to displace proteins in ion exchange systems.

Low molecular mass displacers have significant operational advantages as compared to large polyelectrolyte displacers. For example, if there is any overlap between the displacer and the protein of interest, these low molecular mass materials can be readily separated from the purified protein during post-displacement processing using standard size-based purification methods (e.g. size exclusion chromatography, ultrafiltration). In addition, the salt-dependent adsorption behavior of these low MW displacers greatly facilitates column regeneration. These displacers have been employed for a wide variety of high resolution separations in ion exchange systems. In addition, the utility of displacement chromatography for the purification of recombinant growth factors, antigenicvaccine proteins and antisense oligonucleotides has also been demonstrated. There are several examples in which displacement chromatography has been applied to the purification of proteins using ion exchange, hydrophobic interaction, as well as reversed-phase chromatography.

Displacement chromatography is well suited for obtaining mg quantities of purified proteins from complex mixtures using standard analytical chromatography columns at the bench scale. It is also particularly well suited for enriching trace components in the feed. Displacement chromatography can be readily carried out using a variety of resin systems including, ion exchange, HIC and RPLC

Two-Dimensional Chromatography

Two-dimensional chromatography represents the most thorough and rigorous approach to evaluation of the proteome. While previously accepted approaches have utilized elution mode chromatographic approaches such as cation exchange to reversed phase HPLC, yields are typically very low requiring analytical sensitivities in the picomolar to femtomolar range. As displacement chromatography offers the advantage of concentration of trace components, two dimensional chromatography utilizing displacement rather than elution mode in the upstream chromatography step represents a potentially powerful tool for analysis of trace components, modifications, and identification of minor expressed components of the proteome.

REFERENCES

1. N. Tugcu . Purification of proteins using displacement chromatography. pp 71-89 in M. Zachariou (Ed.) Methods in Molecular Biology: Vol 421 Affinity Chromatography: Methods and Protocols. 2nd edition. Humana Press, Totowa NJ.
2. A. Tiselius. Displacement development in adsorption analysis. Ark. Kemi. Mineral Geol. 16A: 1–18 (1943).
3. G. T. Seaborg. The Transuranium Elements. Science 104(2704):379-386 (1946).
4. J. Frenz and C.S. Horvath. High performance displacement chromatography. pp 212-314 in C. Horvath (Ed.) High Performance Liquid Chromatography-advances and perspectives. Vol. 5, Academic Press, San Diego, CA.

5. C.S. Horvath, A. Nahum, and J. Frenz. High performance displacement chromatography. J. Chromatogr. 218, 365–393(1981).

6. Little, Charles Displacement Chromatography Comes of Age. An indispensable tool for the purification of biomacromolecules Archived February 22, 2011, at the Wayback Machine. *Chromatography Techniques* (Date of original not given on website but Google shows Nov 15, 2008; accessed Feb. 2011)

7. Ford, J.C. "Displacement chromatography" in Jack Cazes, *Encyclopedia of Chromatography*, pp255-257 Marcel Dekker 2001

8. Helfferich, F. *Ion Exchange* McGraw Hill, NY, 1962

9. Buchanan, D.C. Preparative isolation of amino acids by carrier displacement chromatography on ion exchange resins *Journal of Biological Chemistry 229*, 211-229, 1957

10. Partridge, S.M., and R.C. Brimley. Displacement chromatography on ion exchange resins. 8. A systematic method for the separation of Amino Acids. Biochemical Journal 51, 628-639, 1952

11. Spedding, F.H., J.E. Powell, and E.J. Wheelwright. The use of copper as the retaining ion in the elution of rare earths with ammonium ethylenediamine tetraacetate solutions *Journal of the American Chemical Society 76*,2557-60, 1954,

12. Phillips, T.R., D.R. Owens, and A.G. Hamlin. Purification of hydrogen isotopes by self-displacement chromatography at low temperatures *Nature 192* 1067-8, 1961

13. S. M. Cramer and G. Jayaraman, Current Opinions in Biotechnology 4: 217-225, (1993)

14. G. Jayaraman, S. Gadam, and S. M. Cramer. J. Chromatogr. A 630:53-68. (1993)

15. G. Jayaraman, Y. Li, J. A. Moore, and S. M. Cramer. J. Chromatogr. A 702:143-155. (1995)

16. A. Kundu, S. Vunnum, G. Jayaraman, and S. M. Cramer. Biotech. and Bioeng. 48: 452-460. (1995)

17. A. Kundu, S. Vunnum, and S. M. Cramer. J. Chromatogr. A, 707:57-67. (1995)

18. A. Kundu, S. Vunnum, and S. M. Cramer. Adsorption 4:3-4. (1998)

19. A. Kundu, K. Barnthouse, and S. M. Cramer. Biotech. and Bioeng., 56:119-129. (1997)

20. KA. Kundu, A. A. Shukla, K. A. Barnthouse, J. Mooreand S. M. Cramer. BioPharm 10 :64. (1997)

21. A. Kundu, and S. M. Cramer. Anal. Biochem., 248:111-116. (1997)

22. A. A. Shukla, K. A. Barnthouse, S. S. Bae, J. A. Moore, and S. M. Cramer.. J. Chromatogr. A 814:1-2. (1998)

23. K. A. Barnthouse, W. Trompeter, R. Jone, P. Inampudi, R.Rupp, and S. M. Cramer. J. Biotechnol. 66:125-136 (1998)

24. A. A. Shukla, R. L. Hopfer, D. N. Chakravarti, E. Bortell, and S. M. Cramer. Biotechnol. Prog. 14: 91-101(1998)

25. N. Tugcu, R. R. Deshmukh, Y. S. Sangvic, J. A. Moored, and S. M. Cramer J. Chromatogr. A 923:65-73(2001)

26. R. Freitag and J. Breier. J. Chromatogr. A 691, 101–112 (1995).

27. N. Tugcu, R. R. Deshmukh, Y. S. Sanghvi, and S. M. Cramer. Reactive and Functional Polymers 54, 37–47(2003).

28. E. Nagele, M. Vollmer, P. Horth, and C. Vad. 2D-LC/MS techniques for the identification of proteins in highly complex mixtures. Expert Reviews in Proteomics. Vol. 1, No. 1, Pages 37-46 (2004).

CHAPTER 10

Capillary Electrophoresis

Chapter Contents

- **Introduction**
- **Instrumentation**
- **Detection**
- **Modes of Separation**
- **Efficiency and Resolution**
- **Applications**
- **Capillary Electrophoresis–Mass Spectrometry**
- **Tandem Mass Spectrometry**
- **Instrumentation**
- **Electrostatic Precipitator**
- **Microfiltration**

Capillary electrophoresis is an analytical technique that separates ions based on their electrophoretic mobility with the use of an applied voltage. The electrophoretic mobility is dependent upon the charge of the molecule, the viscosity, and the atom's radius. The rate at which the particle moves is directly proportional to the applied electric field--the greater the field strength, the faster the mobility. Neutral species are not affected, only ions move with the electric field. If two ions are the same size, the one with greater charge will move the fastest. For ions of the same charge, the smaller particle has less friction and overall faster migration rate. Capillary electrophoresis is used most predominately because it gives faster results and provides high resolution separation. It is a useful technique because there is a large range of detection methods available.

INTRODUCTION

Capillary electrophoresis (**CE**) is a family of electrokinetic separation methods performed in submillimeter diameter capillaries and in micro- and nanofluidic channels. Very often, CE refers to capillary zone electrophoresis (CZE), but other electrophoretic techniques including capillary gel electrophoresis (CGE), capillary isoelectric focusing (CIEF), capillary isotachophoresis and micellar electrokinetic chromatography (MEKC) belong also to this class of methods. In CE methods, analytes migrate through electrolyte solutions under the influence of an electric field. Analytes can be separated according to ionic mobility and/or partitioning into an alternate phase via Non-covalent interactions. Additionally, analytes may be concentrated or "focused" by means of gradients in conductivity and pH.

INSTRUMENTATION

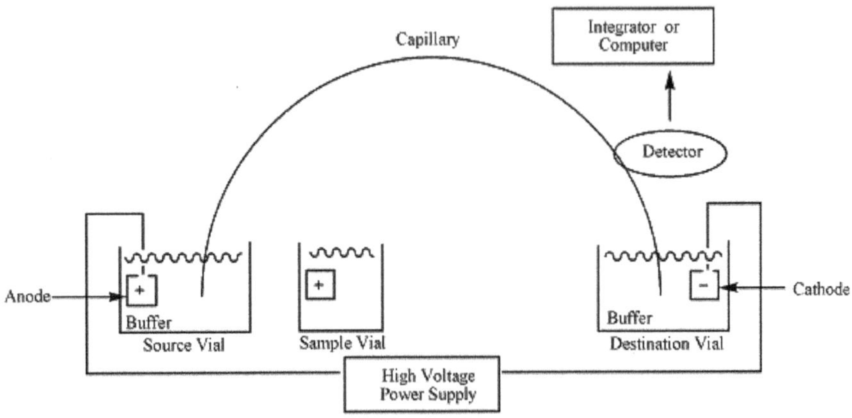

Figure 1: Diagram of capillary electrophoresis system.

The instrumentation needed to perform capillary electrophoresis is relatively simple. A basic schematic of a capillary electrophoresis system is shown in *figure 1*. The system's main components are a sample vial, source and destination vials, a capillary, electrodes, a high voltage power supply, a detector, and a data output and handling device. The source vial, destination vial and capillary are filled with an electrolyte such as an aqueous buffer solution. To introduce the sample, the capillary inlet is placed into a vial containing the sample. Sample is introduced into the capillary via capillary action, pressure, siphoning, or electrokinetically, and the capillary is then returned to the source vial. The migration of the analytes is initiated by an electric field that is applied between the source and destination vials and is supplied to the electrodes by the high-voltage power supply. In the most common mode of CE, all ions, positive or negative, are pulled through the capillary in the same direction by electroosmotic flow. The analytes separate as they migrate due to their electrophoretic mobility, and are detected near the outlet end of the capillary. The output of the detector is sent to a data output and handling device such as an integrator or computer. The data is then displayed as an electropherogram, which reports detector response as a function of time. Separated chemical compounds appear as peaks with different retention times in an electropherogram. Capillary electrophoresis was first combined with mass spectrometry by Richard D. Smith and coworkers, and provides extremely high sensitivity for the analysis of very small sample sizes. Despite the very small sample sizes (typically only a few nanoliters of liquid are introduced into the capillary), high sensitivity and sharp peaks are achieved in part due to injection strategies that result in concentration of analytes into a narrow zone near the inlet of the capillary. This is achieved in either pressure or electrokinetic injections simply by suspending the sample in a buffer of lower conductivity (*e.g.* lower salt concentration) than the running buffer. A process called field-amplified sample stacking (a form of isotachophoresis) results in concentration of analyte in a narrow zone at the boundary between the low-conductivity sample and the higher-conductivity running buffer.

To achieve greater sample throughput, instruments with arrays of capillaries are used to analyze many samples simultaneously. Such capillary array electrophoresis (CAE) instruments with 16 or 96 capillaries are used for medium- to high-throughput capillary DNA sequencing, and the inlet ends of the capillaries are arrayed spatially to accept samples directly from SBS-standard footprint 96-well plates. Certain aspects of the instrumentation (such as detection) are necessarily more complex than for a single-capillary system, but the fundamental principles of design and operation are similar to those shown in Figure 1.

DETECTION

Separation by capillary electrophoresis can be detected by several detection devices. The majority of commercial systems use UV or UV-Vis absorbance as their primary mode of detection. In these systems, a section of the capillary itself is used as the detection cell. The use of on-tube detection enables detection of separated analytes with no loss of resolution. In general, capillaries used in capillary electrophoresis are coated with a polymer (frequently polyimide or Teflon) for increased flexibility. The portion of the capillary used for UV detection, however, must be optically transparent. For polyimide-coated capillaries, a segment of the coating is typically burned or scraped off to provide a bare window several millimeters long. This bare section of capillary can break easily, and capillaries with transparent coatings are available to increase the stability of the cell window. The path length of the detection cell in capillary electrophoresis (~ 50 micrometers) is far less than that of a traditional UV cell (~ 1 cm). According to the Beer-Lambert law, the sensitivity of the detector is proportional to the path length of the cell. To improve the sensitivity, the path length can be increased, though this results in a loss of resolution. The capillary tube itself can be expanded at the detection point, creating a "bubble cell" with a longer path length or additional tubing can be added at the detection point as shown in *figure 2*. Both of these methods, however, will decrease the resolution of the separation. Post-column detection utilizing a sheath flow configuration has also been described.

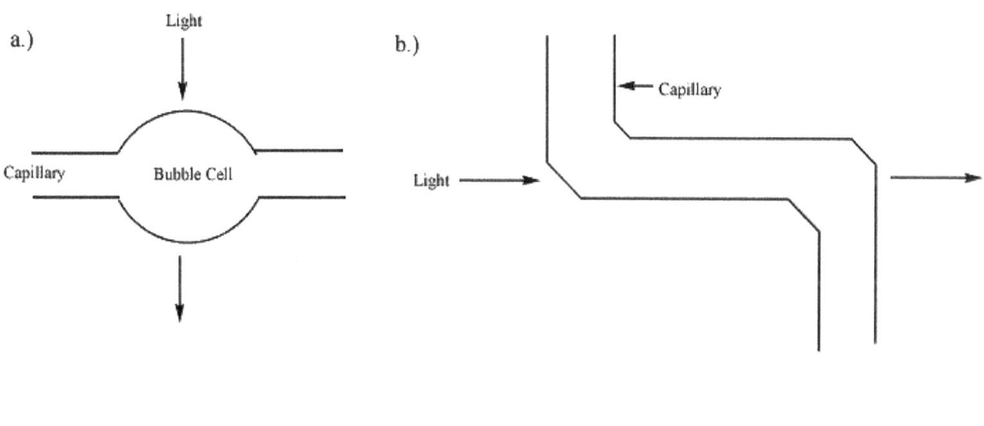

Figure 2: Techniques for increasing the pathlength of the capillary: a.) a bubble cell and b.) a z-cell (additional tubing).

Fluorescence detection can also be used in capillary electrophoresis for samples that naturally fluoresce or are chemically modified to contain fluorescent tags. This mode of detection offers high sensitivity and improved selectivity for these samples, but cannot be utilized for samples that do not fluoresce. Numerous labeling strategies are used to create fluorescent derivatives or conjugates of non-fluorescent molecules, including proteins and DNA. The set-up for fluorescence detection in a capillary electrophoresis system can be complicated. The method requires that the light beam be focused on the capillary, which can be difficult for many light sources. Laser-induced fluorescence has been used in CE systems with detection limits as low as 10^{-18} to 10^{-21} mol. The sensitivity of the technique is attributed to the high intensity of the incident light and the ability to accurately focus the light on the capillary. Multi-color fluorescence detection can be achieved by including multiple dichroic mirrors and bandpass filters to separate the fluorescence emission amongst multiple detectors (*e.g.*, photomultiplier tubes), or by using a prism or grating to project spectrally resolved fluorescence emission onto a position-sensitive detector such as a CCD array. CE systems with 4- and 5-color LIF detection systems are used routinely for capillary DNA sequencing and genotyping ("DNA fingerprinting") applications.

In order to obtain the identity of sample components, capillary electrophoresis can be directly coupled with mass spectrometers or Surface Enhanced Raman Spectroscopy (SERS). In most systems, the capillary outlet is introduced into an ion source that utilizes electrospray ionization (ESI). The resulting ions are then analyzed by the mass spectrometer. This set-up requires volatile buffer solutions, which will affect the range of separation modes that can be employed and the degree of resolution that can be achieved. The measurement and analysis are mostly done with a specialized gel analysis software.

For CE-SERS, capillary electrophoresis eluants can be deposited onto a SERS-active substrate. Analyte retention times can be translated into spatial distance by moving the SERS-active substrate at a constant rate during capillary electrophoresis. This allows the subsequent spectroscopic technique to be applied to specific eluants for identification with high sensitivity. SERS-active substrates can be chosen that do not interfere with the spectrum of the analytes.

MODES OF SEPARATION

The separation of compounds by capillary electrophoresis is dependent on the differential migration of analytes in an applied electric field. The electrophoretic migration velocity (u_p) of an analyte toward the electrode of opposite charge is:

$$u_p = \mu_p E$$

The electrophoretic mobility can be determined experimentally from the migration time and the field strength:

$$\mu_p = \left(\frac{L}{t_r}\right)\left(\frac{L_t}{V}\right)$$

where L is the distance from the inlet to the detection point, t_r is the time required for the analyte to reach the detection point (migration time), V is the applied voltage (field strength), and L_t is the total length of the capillary. Since only charged ions are affected by the electric field, neutral analytes are poorly separated by capillary electrophoresis.

The velocity of migration of an analyte in capillary electrophoresis will also depend upon the rate of electroosmotic flow (EOF) of the buffer solution. In a typical system, the electroosmotic flow is directed toward the negatively charged cathode so that the buffer flows through the capillary from the source vial to the destination vial. Separated by differing electrophoretic mobilities, analytes migrate toward the electrode of opposite charge. As a result, negatively charged analytes are attracted to the positively charged anode, counter to the EOF, while positively charged analytes are attracted to the cathode, in agreement with the EOF as depicted in *figure 3*.

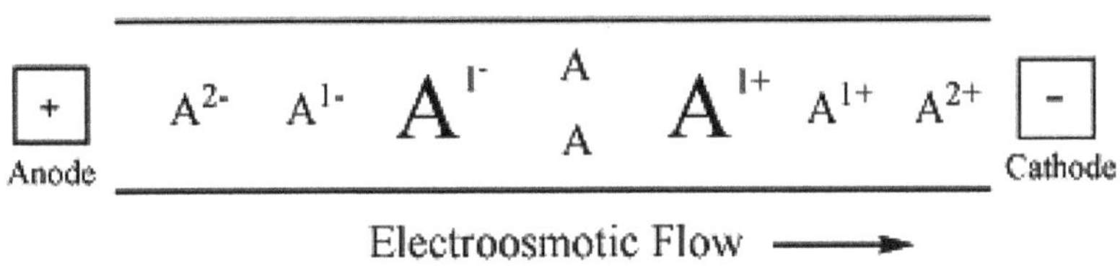

Figure 3: Diagram of the separation of charged and neutral analytes (A) according to their respective electrophoretic and electroosmotic flow mobilities.

The velocity of the electroosmotic flow, u_o can be written as:

$$u_o = \mu_o E$$

where μ_o is the electroosmotic mobility, which is defined as:

$$\mu_o = \frac{\epsilon \zeta}{\eta}$$

where ζ is the zeta potential of the capillary wall, and ϵ is the relative permittivity of the buffer solution. Experimentally, the electroosmotic mobility can be determined by measuring the retention time of a neutral analyte.The velocity u of an analyte in an electric field can then be defined as:

$$u_p + u_o = (\mu_p + \mu_o)E$$

Since the electroosmotic flow of the buffer solution is generally greater than that of the electrophoretic mobility of the analytes, all analytes are carried along with the buffer solution toward the cathode. Even small, triply charged anions can be redirected to the cathode by the relatively powerful EOF of the buffer solution. Negatively charged analytes are retained longer in the capilliary due to their conflicting electrophoretic mobilities. The order of migration seen by the detector is shown in *figure 3*: small multiply charged cations migrate quickly and small multiply charged anions are retained strongly.

Electroosmotic flow is observed when an electric field is applied to a solution in a capillary that has fixed charges on its interior wall. Charge is accumulated on the inner surface of a capillary when a buffer solution is placed inside the capillary. In a fused-silica capillary, silanol (Si-OH) groups attached to the interior wall of the capillary are ionized to negatively charged silanoate ($Si-O^-$) groups at pH values greater than three. The ionization of the capillary wall can be enhanced by first running a basic solution, such as NaOH or KOH through the capillary prior to introducing the buffer solution. Attracted to the negatively charged silanoate groups, the positively charged cations of the buffer solution will form two inner layers of cations (called the diffuse double layer or the electrical double layer) on the capillary wall as shown in *figure 4*. The first layer is referred to as the fixed layer because it is held tightly to the silanoate groups. The outer layer, called the mobile layer, is farther from the silanoate groups. The mobile cation layer is pulled in the direction of the negatively charged cathode when an electric field is applied. Since these cations are solvated, the bulk buffer solution migrates with the mobile layer, causing the electroosmotic flow of the buffer solution. Other capillaries including Teflon capillaries also exhibit electroosmotic flow. The EOF of these capillaries is probably the result of adsorption of the electrically charged ions of the buffer onto the capillary walls. The rate of EOF is dependent on the field strength and the charge density of the capillary wall. The wall's charge density is proportional to the pH of the buffer solution. The electroosmotic flow will increase with pH until all of the available silanols lining the wall of the capillary are fully ionized.

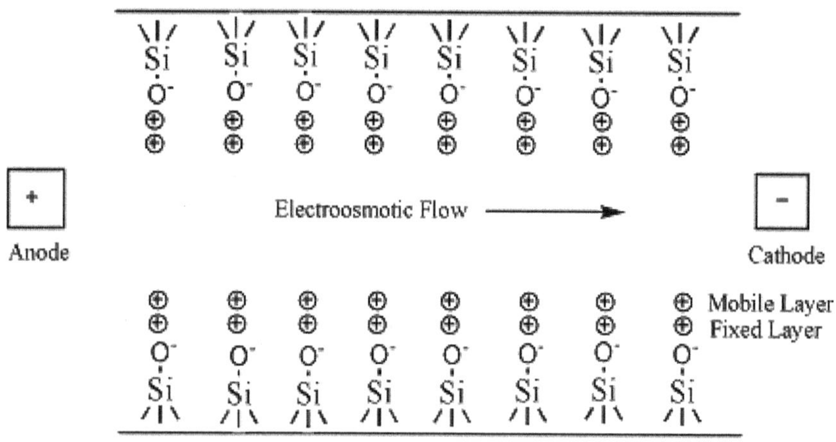

Figure 4: Depiction of the interior of a fused-silica gel capillary in the presence of a buffer solution.

In certain situations where strong electroosomotic flow toward the cathode is undesirable, the inner surface of the capillary can be coated with polymers, surfactants, or small molecules to reduce electroosmosis to very low levels, restoring the normal direction of migration (anions towar the anode, cations toward the cathode). CE instrumentation typically includes power supplies with reversible polarity, allowing the same instrument to be used in "normal" mode (with EOF and detection near the cathodic end of the capillary) and "reverse" mode (with EOF suppressed or reversed, and detection near the anodic end of the capillary). One of the most common approaches to suppressing EOF, reported by Stellan Hjertén in 1985, is to create a covalently attached layer of linear polyacrylamide. The silica surface of the capillary is first modified with a silane reagent bearing a polymerizable vinyl group (*e.g.* 3-methacryloxypropyltrimethoxysilane), followed by introduction of acrylamide monomer and a free radical initiator. The acrylamide is polymerized *in situ*, forming long linear chains, some of which are covalently attached to the wall-bound silane reagent. Numerous other strategies for covalent modification of capillary surfaces exist. Dynamic or adsorbed coatings (which can include polymers or small molecules) are also common. For example, in capillary sequencing of DNA, the sieving polymer (typically polydimethylacrylamide) suppresses electroosmotic flow to very low levels. A variety of dynamic capillary coating agents are commercially available to modify, suppress, or reverse the direction of electroosmotic flow. Besides modulating electroosmotic flow, capillary wall coatings can also serve the purpose of reducing interactions between "sticky" analytes (such as proteins) and the capillary wall. Such wall-analyte interactions, if severe, manifest as reduced peak efficiency, asymmetric (tailing) peaks, or even complete loss of analyte to the capillary wall.

EFFICIENCY AND RESOLUTION

The number of theoretical plates, or separation efficiency, in capillary electrophoresis is given by:

$$N = \frac{\mu V}{2D_m}$$

where N is the number of theoretical plates, μ is the apparent mobility in the separation medium and D_m is the diffusion coefficient of the analyte. According to this equation, the efficiency of separation is only limited by diffusion and is proportional to the strength of the electric field, although practical considerations limit the strength of the electric field to several hundred volts per centimeter. Application of very high potentials (>20-30 kV) may lead to arcing or breakdown of the capillary. Further, application of strong electric fields leads to resistive heating (Joule heating) of the buffer in the capillary. At sufficiently high field strengths, this heating is strong enough that radial temperature gradients can develop within the capillary. Since electrophoretic mobility of ions is generally temperature-dependent (due to both temperature-dependent ionization and solvent viscosity effects), a non-uniform temperature profile results in variation of electrophoretic mobility across the capillary, and a loss of resolution. The onset of significant Joule heating can be determined by constructing an "Ohm's Law plot", wherein the current through the capillary is measured as a function of applied potential. At low fields, the current is proportional to the applied potential (Ohm's Law), whereas at higher fields the current deviates from the straight line as heating results in decreased resistance of the buffer. The best resolution is typically obtained at the maximum field strength for which Joule heating is insignificant (*i.e.* near the boundary between the linear and nonlinear regimes of the Ohm's Law plot). Generally capillaries of smaller inner diameter support use of higher field strengths, due to improved heat dissipation and smaller thermal gradients relative to larger capillaries, but with the drawbacks of lower sensitivity in absorbance detection due to shorter path length, and greater difficulty in introducing buffer and sample into the capillary (small capillaries require greater pressure and/or longer times to force fluids through the capillary).

The efficiency of capillary electrophoresis separations is typically much higher than the efficiency of other separation techniques like HPLC. Unlike HPLC, in capillary electrophoresis there is no mass transfer between phases. In addition, the flow profile in EOF-driven systems is flat, rather than the rounded laminar flow profile characteristic of the pressure-driven flow in chromatography columns as shown in *figure 5*. As a result, EOF does not significantly contribute to band broadening as in pressure-driven chromatography. Capillary electrophoresis separations can have several hundred thousand theoretical plates.

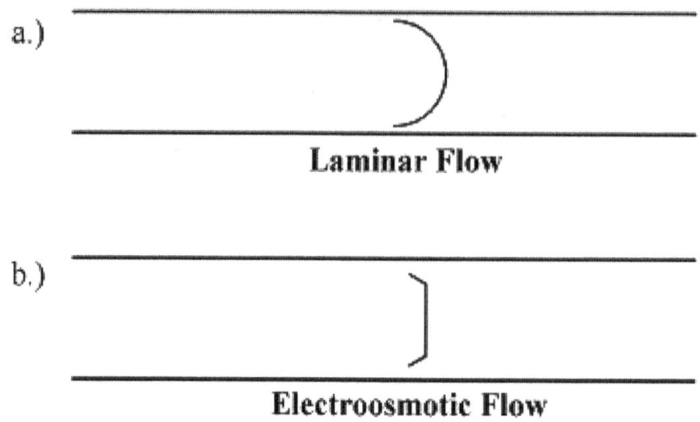

Figure 5: Flow profiles of laminar and electroosmotic flow.

The resolution (R_s) of capillary electrophoresis separations can be written as:

$$R_s = \frac{1}{4}\left(\frac{\Delta\mu_p\sqrt{N}}{\mu_p + \mu_o}\right)$$

According to this equation, maximum resolution is reached when the electrophoretic and electroosmotic mobilities are similar in magnitude and opposite in sign. In addition, it can be seen that high resolution requires lower velocity and, correspondingly, increased analysis time.

Besides diffusion and Joule heating (discussed above), factors that may decrease the resolution in capillary electrophoresis from the theoretical limits in the above equation include, but are not limited to, the finite widths of the injection plug and detection window; interactions between the analyte and the capillary wall; instrumental non-idealities such as a slight difference in height of the fluid reservoirs leading to siphoning; irregularities in the electric field due to, *e.g.*, imperfectly cut capillary ends; depletion of buffering capacity in the reservoirs; and electrodispersion (when an analyte has higher conductivity than the background electrolyte). Identifying and minimizing the numerous sources of band broadening is key to successful method development in capillary electrophoresis, with the objective of approaching as close as possible to the ideal of diffusion-limited resolution.

APPLICATIONS

Capillary electrophoresis may be used for the simultaneous determination of the ions NH_4^+, Na^+, K^+, Mg^{2+} and Ca^{2+} in saliva.

CAPILLARY ELECTROPHORESIS–MASS SPECTROMETRY

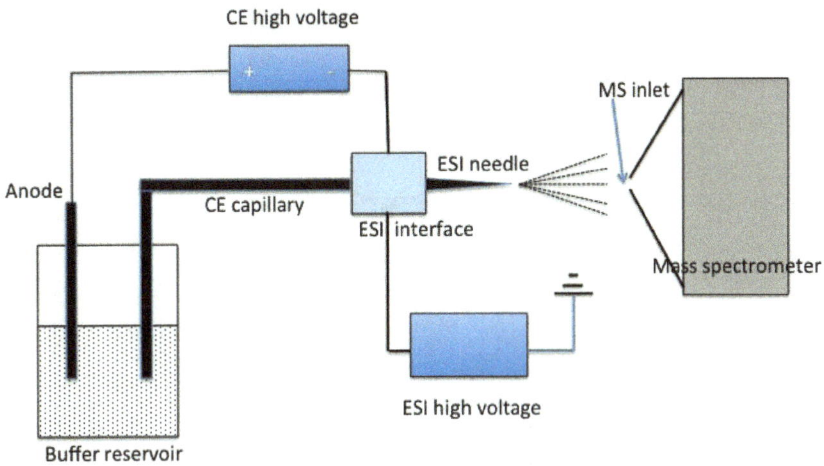

Schematic Diagram of CE/ESI-MS.

Capillary electrophoresis–mass spectrometry (CE-MS) is an analytical chemistry technique formed by the combination of the liquid separation process of capillary electrophoresis with mass spectrometry. CE-MS combines advantages of both CE and MS to provide high separation efficiency and molecular mass information in a single analysis. It has high resolving power and sensitivity, requires minimal volume and can analyze at high speed. Ions are typically formed by electrospray ionization, but they can also be formed by matrix-assisted laser desorption/ionization or other ionization techniques. It has applications in basic research in proteomics and quantitative analysis of biomolecules as well as in clinical medicine.

Since its introduction in 1987, new developments and application has made CE-MS powerful separation and identification technique. Use of CE-MS has increased for protein and peptides analysis and other biomolecules. Understanding of CE, the interface setup, ionization technique and mass detection system is important to tackle problems while coupling capillary electrophoresis to mass spectrometry.

The original interface between capillary zone electrophoresis and mass spectrometry was developed in 1987 by Richard D. Smith and coworkers at Pacific Northwest National Laboratory, and who also later were involved in development of interfaces with other CE variants, including capillary isotachophoresis and capillary isoelectric focusing.

Interfacing CE with MS

Capillary electrophoresis is a separation technique which uses high electric field to produce electroosmotic flow for separation of ions. Analytes migrate from one end of capillary to other based on their charge, viscosity and size. Higher the electric field, greater is the mobility. Mass spectrometry is an analytical technique that identifies chemical species depending on their mass-to-charge ratio. During the process, an ion source will convert molecules coming from CE to ions that can then be manipulated using electric and magnetic field. The separated ions are then measured using a detector. The major problem faced when coupling CE to MS arises due to insufficient understanding of fundamental processes when two techniques are interfaced. The separation and detection of analytes can be improved with better interface. CE has been coupled to MS using various ionization techniques like FAB, ESI, MALDI, APCI and DESI. The most used ionization technique is ESI.

Electrospray Ionization Interface

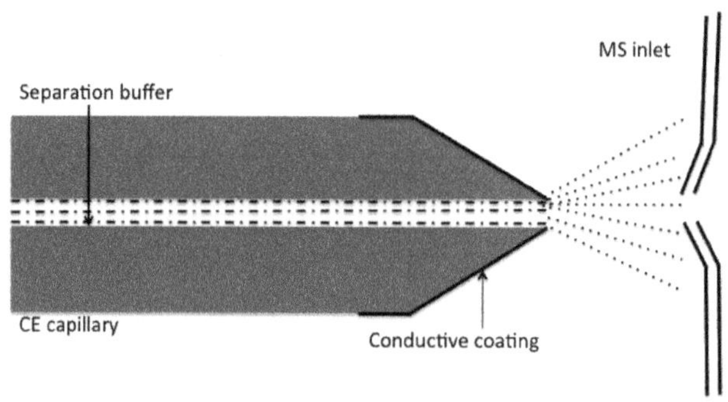

Sheathless Interface.

The first CE-MS interface had cathode end of CE capillary terminated within a stainless steel capillary. An electrical contact was made at that point completing the circuit and initiating the electrospray. This interface system had few drawbacks like mismatch in the flow rates of two systems. Since then, interface system has been improved to have continuous flow rate and good electrical contact. At present, three types of interface system exist for CE/ESI-MS which are discussed briefly.

Sheathless Interface

CE capillary is coupled directly to ionization source in sheathless interface system. The capillary is coated with conducting metal or gold or platinum wire is inserted into CE capillary to establish suitable electrical connection. Since no sheath liquid is used, the system has high sensitivity, low flow rates and minimum background. However, right choice of buffer solution has to be made which is suitable for both CE separation and ESI operation.

Sheath Flow Interface

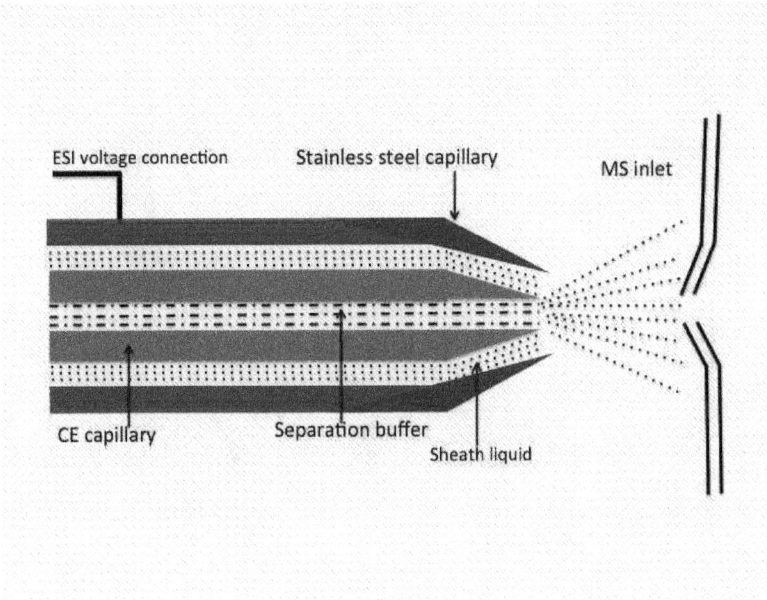

Sheath Flow Interface.

It is most widely used interface system. The electrical connection is established when the CE separation liquid is mixed with sheath liquid flowing coaxially in a metal capillary tubing. Commonly used sheath liquid is 1:1 mixture of water-methanol with 0.1% acetic acid or formic acid. The system is more reliable and has wide selection range of separation electrolyte. There might be some decrease in sensitivity due to sheath liquid.

Liquid Junction Interface

This technique uses a stainless steel tee to mix separation electrolyte from CE capillary with make-up liquid. The CE capillary and ESI needle are inserted through opposite sides of the tee and a narrow gap is maintained. The electrical contact is established by makeup liquid surrounding the junction between two capillaries. This system is easy to operate. However, the sensitivity is reduced and the mixing of two liquids could degrade separation.

Continuous- Flow Fast Atom Bombardment

Miniard et al. first coupled FAB to CE in 1988. The main problem of this interface is the flow rate between the two systems. The CF-FAB needs very high flow rate but CE need very minimal flow rate for better separation. So a make up flow is introduced using the designs discussed above: sheath flow or liquid junction. CF/FAB-MS is not commonly used as CE/ESI-MS, as FAB is rarely used anymore.

Coupling CE with MALDI-MS

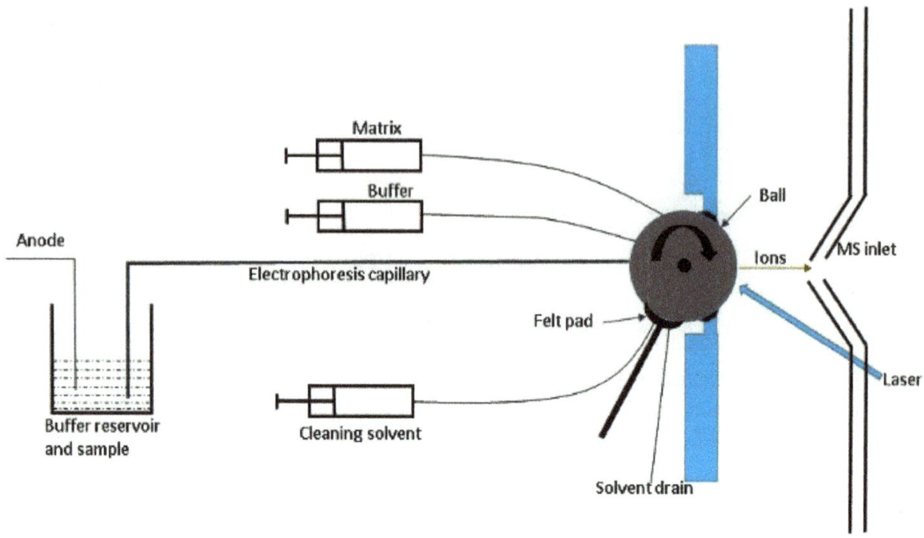

Schematic diagram of online CE-MALDI-MS.

Off-line coupling of CE to MALDI, the CE effluent could be sprayed or added drop wise on MALDI target plate then dried and analyzed by MS. For online coupling, a moving target with continuous contact to CE capillary end is required. The moving target takes analytes into MS where it is desorbed and ionized. Musyimi et al. developed a new technique where rotating ball was used to transfer CE to MS. The sample from CE is mixed with matrix coming though another capillary. As the ball rotates the sample is dried before it reaches ionization region. This technique has high sensitivity since no makeup fluid is used.

Applications

CE-MS ability to separate analytes present in extremely low concentration with high efficiency at high speed has made it applicable in all fields of science. CE-MS has been used for bioanalytical, pharmaceuticals, environmental and forensic application. The major application of CE-MS has been for biological studies, mostly for protein and peptide analysis. Along with that, it is used often for routine analysis of pharmaceutical drugs. There are number of studies reporting characterization of mixtures of peptides and proteins.CE-MS can be used for routine clinical checkup. Body fluids like blood and urine have been analyzed with CE-MS to identify biomarkers for renal diseases and cancer.

TANDEM MASS SPECTROMETRY

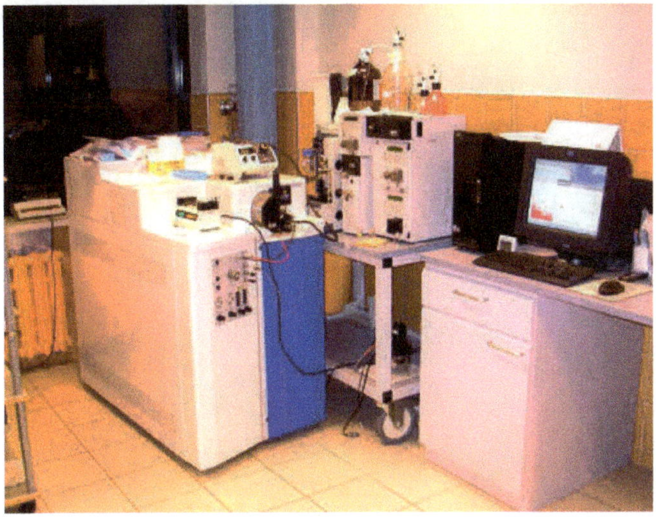

A quadrupole time-of-flight hybrid tandem mass spectrometer.

Tandem mass spectrometry, also known as **MS/MS** or **MS²**, involves multiple steps of mass spectrometry selection, with some form of fragmentation occurring in between the stages. In a tandem mass spectrometer, ions are formed in the ion source and separated by mass-to-charge ratio in the first stage of mass spectrometry (MS1). Ions of a particular mass-to-charge ratio (precursor ions) are selected and fragment ions (product ions) are created by collision-induced dissociation, ion-molecule reaction, photodissociation, or other process. The resulting ions are then separated and detected in a second stage of mass spectrometry (MS2).

INSTRUMENTATION

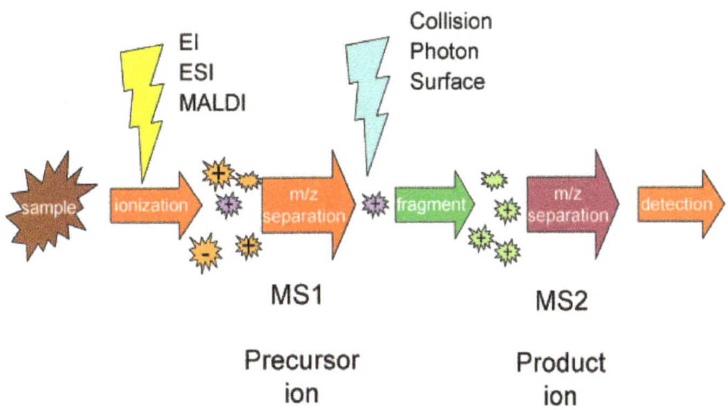

Schematic of tandem mass spectrometry.

For tandem mass spectrometry in space, the different elements are often noted in shorthand. Multiple stages of mass analysis separation can be accomplished with individual mass spectrometer elements separated in space or using a single mass spectrometer with the MS steps separated in time.

Tandem in Space

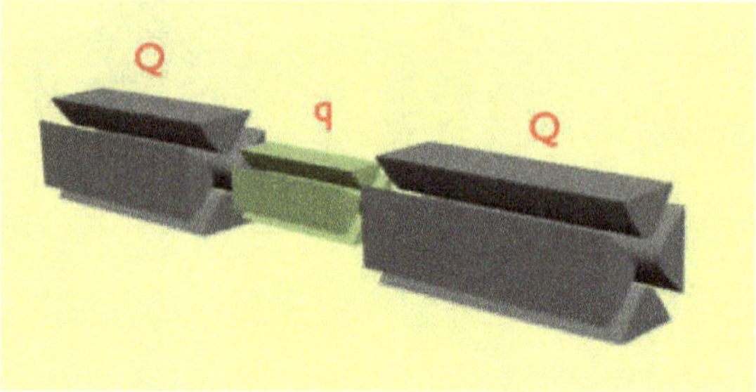

Triple quadrupole diagram; and example of tandem mass spectrometry in space.

In tandem mass spectrometry *in space*, the separation elements are physically separated and distinct, although there is a physical connection between the elements to maintain high vacuum. These elements can be sectors, transmission quadrupole, or time-of-flight. When using multiple quadrupoles, they can act as both mass analyzers and collision chambers.

- **Q** – Quadrupole mass analyzer
- **q** – Radio frequency collision quadrupole
- **TOF** – Time-of-flight mass analyzer
- **B** – Magnetic sector
- **E** – Electric sector
- The notation can be combined to indicate various hybrid instrument, for example
- **QqQ** – Triple quadrupole mass spectrometer
- **QTOF** – Quadrupole time-of-flight mass spectrometer (also QqTOF)
- **BEBE** – Four-sector (reverse geometry) mass spectrometer

Tandem in time

By doing tandem mass spectrometry in time, the separation is accomplished with ions trapped in the same place, with multiple separation steps taking place over time. A quadrupole ion trap or Fourier transform ion cyclotron resonance (FTICR) instrument can be used for such an analysis. Trapping instruments can perform multiple steps of analysis, which is sometimes referred to as MSn (MS to the n). Often the number of steps, n, is not indicated, but occasionally the value is specified; for example MS3 indicates three stages of separation. Tandem in time MS instruments do not use the modes described next, but typically collect all of the information from a precursor ion scan and a parent ion scan of the entire spectrum. Each instrumental configuration utilizes a unique mode of mass identification.

An ion trap mass spectrometer is an example of a tandem mass spectrometry in time instrument.

Tandem in Space MS/MS Modes

When tandem MS is performed with an in space design, the instrument must operate in one of a variety of modes. There are a number of different tandem MS/MS experimental setups and each mode has its own applications and provides different information. Tandem MS in space uses the coupling of two instrument components which measure the same mass spectrum range but with a controlled fractionation between them in space, while tandem MS in time involves the use of an ion trap.

There are four main scan experiments possible using MS/MS: precursor ion scan, product ion scan, neutral loss scan, and selected reaction monitoring.

For a precursor ion scan, the product ion is selected in the second mass analyzer, and the precursor masses are scanned in the first mass analyzer. Note that precursor ion is synonymous with parent ion and product ion with daughter ion; however the use of these anthropomorphic terms is discouraged.

In a product ion scan, a precursor ion is selected in the first stage, allowed to fragment and then all resultant masses are scanned in the second mass analyzer and detected in the detector that is positioned after the second mass analyzer. This experiment is commonly performed to identify transitions used for quantification by tandem MS.

In a neutral loss scan, the first mass analyzer scans all the masses. The second mass analyzer also scans, but at a set offset from the first mass analyzer. This offset corresponds to a neutral loss that is commonly observed for the class of compounds. In a constant-neutral-loss scan, all precursors that undergo the loss of a specified common neutral are monitored. To obtain this information, both mass analyzers are scanned simultaneously, but with a mass offset that correlates with the mass of the specified neutral. Similar to the precursor-ion scan, this technique is also useful in the selective identification of closely related class of compounds in a mixture.

In selected reaction monitoring, both mass analyzers are set to a selected mass. This mode is analogous to selected ion monitoring for MS experiments. A selective analysis mode, which can increase sensitivity.

Fragmentation

Fragmentation of gas-phase ions is essential to tandem mass spectrometry and occurs between different stages of mass analysis. There are many methods used to fragment the ions and these can result in different types of fragmentation and thus different information about the structure and composition of the molecule.

In-Source Fragmentation

Often, the ionization process is sufficiently violent to leave the resulting ions with sufficient internal energy to fragment within the mass spectrometer. If the product ions persist in their non-equilibrium state for a moderate amount of time before auto-dissociation this process is called metastable fragmentation. Nozzle-skimmer fragmentation refers to the purposeful induction of in-source fragmentation by increasing the nozzle-skimmer potential on usually electrospray based instruments. Although in-source fragmentation allows for fragmentation analysis, it is not technically tandem mass spectrometry unless metastable ions are mass analyzed or selected before auto-dissociation and a second stage of analysis is performed on the resulting fragments. In-source fragmentation is often used in addition to tandem mass spectrometry (with post-source fragmentation) to allow for two steps of fragmentation in a pseudo MS^3-type of experiment.

Collision-Induced Dissociation

Post-source fragmentation is most often what is being used in a tandem mass spectrometry experiment. Energy can also be added to the ions, which are usually already vibrationally excited, through post-source collisions with neutral atoms or molecules, the absorption of radiation, or the transfer or capture of an electron by a multiply charged ion. Collision-induced dissociation (CID), also called collisionally activated dissociation (CAD), involves the collision of an ion with a neutral atom or molecule in the gas phase and subsequent dissociation of the ion. For example, consider

$$AB^+ + M \longrightarrow A + B^+ + M$$

where the ion AB^+ collides with the neutral species M and subsequently breaks apart. The details of this process are described by collision theory.

Higher-energy collisional dissociation (HCD) is a CID technique specific to orbitrap mass spectrometers in which fragmentation takes place external to the trap.

Electron Capture and Transfer Methods

The energy released when an electron is transferred to or captured by a multiply charged ion can induce fragmentation.

Electron Capture Dissociation

If an electron is added to a multiply charged positive ion, the Coulomb energy is liberated. Adding a free electron is called electron capture dissociation (ECD), and is represented by

$$[M + nH]^{n+} + e^{-} \longrightarrow \left[[M + (n-1)H]^{(n-1)+}\right]^{*} \longrightarrow \text{fragments}$$

for a multiply protonated molecule M.

Electron Transfer Dissociation

Adding an electron through an ion-ion reaction is called electron transfer dissociation (ETD). Similar to electron-capture dissociation, ETD induces fragmentation of cations (e.g. peptides or proteins) by transferring electrons to them. It was invented by Donald F. Hunt, Joshua Coon, John E. P. Syka and Jarrod Marto at the University of Virginia.

ETD does not use free electrons but employs radical anions (e.g. anthracene or azobenzene) for this purpose:

$$[M + nH]^{n+} + A^{-} \longrightarrow \left[[M + (n-1)H]^{(n-1)+}\right]^{*} + A \longrightarrow \text{fragments}$$

where A is the anion.

ETD cleaves randomly along the peptide backbone (c and z ions) while side chains and modifications such as phosphorylation are left intact. The technique only works well for higher charge state ions (z>2), however relative to collision-induced dissociation (CID), ETD is advantageous for the fragmentation of longer peptides or even entire proteins. This makes the technique important for top-down proteomics. Much like ECD, ETD is effective for peptides with modifications such as phosphorylation.

Electron-transfer and higher-energy collision dissociation (EThcD) is a combination ETD and HCD where the peptide precursor is initially subjected to an ion/ion reaction with fluoranthene anions in a linear ion trap, which generates c- and z-ions. In the second step HCD all-ion fragmentation is applied to all ETD derived ions to generate b- and y- ions prior to final analysis in the orbitrap analyzer. This method employs dual fragmentation to generate ion- and thus data-rich MS/MS spectra for peptide sequencing and PTM localization.

Negative Electron Transfer Dissociation

Fragmentation can also occur with a deprotonated species, in which an electron is transferred from the specie to an cationic reagent in a negative electron transfer dissociation (NETD):

$$[M - nH]^{n-} + A^{+} \longrightarrow \left[[M - nH]^{(n+1)-}\right]^{*} + A \longrightarrow \text{fragments}$$

Following this transfer event, the electron deficient anion undergoes internal rearrangement and fragments. NETD is the ion/ion analogue of electron-detachment dissociation (EDD).

NETD is compatible with fragmenting peptide and proteins along the backbone at the C_{α}-C bond. The resulting fragments are usually a•- and x-type product ions.

Electron-Detachment Dissociation

Electron-detachment dissociation (EDD) is a method for fragmenting anionic species in mass spectrometry. It serves as a negative counter mode to electron capture dissociation. Negatively charged ions are activated by irradiation with electrons of moderate kinetic energy. The result is ejection of electrons from the parent ionic molecule, which causes dissociation via recombination.

Charge Transfer Dissociation

Reaction between positively charged peptides and cationic reagents, also known as charge transfer dissociation (CTD), has recently been demonstrated as an alternative high-energy fragmentation pathway for low-charge state (1+ or 2+) peptides. The proposed mechanism of CTD using helium cations as the reagent is:

$$[\mathrm{M+H}]^{1+} + \mathrm{He}^{+} \longrightarrow \left[[\mathrm{M+H}]^{2+}\right]^{*} + \mathrm{He}^{0} \longrightarrow \textbf{fragments}$$

Initial reports are that CTD causes backbone C_α-C bond cleavage of peptides and provides a•- and x-type product ions.

Photodissociation

The energy required for dissociation can be added by photon absorption, resulting in ion photodissociation and represented by

$$\mathrm{AB}^{+} + h\nu \longrightarrow \mathrm{A} + \mathrm{B}^{+}$$

where $h\nu$ represents the photon absorbed by the ion. Ultraviolet lasers can be used, but can lead to excessive fragmentation of biomolecules.

Infrared multiphoton dissociation

Infrared photons will heat the ions and cause dissociation if enough of them are absorbed. This process is called infrared multiphoton dissociation (IRMPD) and is often accomplished with a carbon dioxide laser and an ion trapping mass spectrometer such as a FTMS.

Blackbody infrared radiative dissociation

Blackbody radiation can be used for photodissociation in a technique known as blackbody infrared radiative dissociation (BIRD). In the BIRD method, the entire mass spectrometer vacuum chamber is heated to create infrared radiation. BIRD uses the light from black body radiation to thermally (vibrationally) excite the ions until a bond breaks. This is similar to infrared multiphoton dissociation with the exception of the source of radiation. This technique is most often used with Fourier transform ion cyclotron resonance mass spectrometers.

Surface Induced Dissociation

With surface-induced dissociation (SID), the fragmentation is a result of the collision of an ion with a surface under high vacuum.

Quantitative Proteomics

Quantitative proteomics is used to determine the relative or absolute amount of proteins in a sample. Several quantitative proteomics methods are based on tandem mass spectrometry. MS/MS has become a benchmark procedure for the structural elucidation of complex biomolecules.

Isobaric tag for Relative and Absolute Quantitation

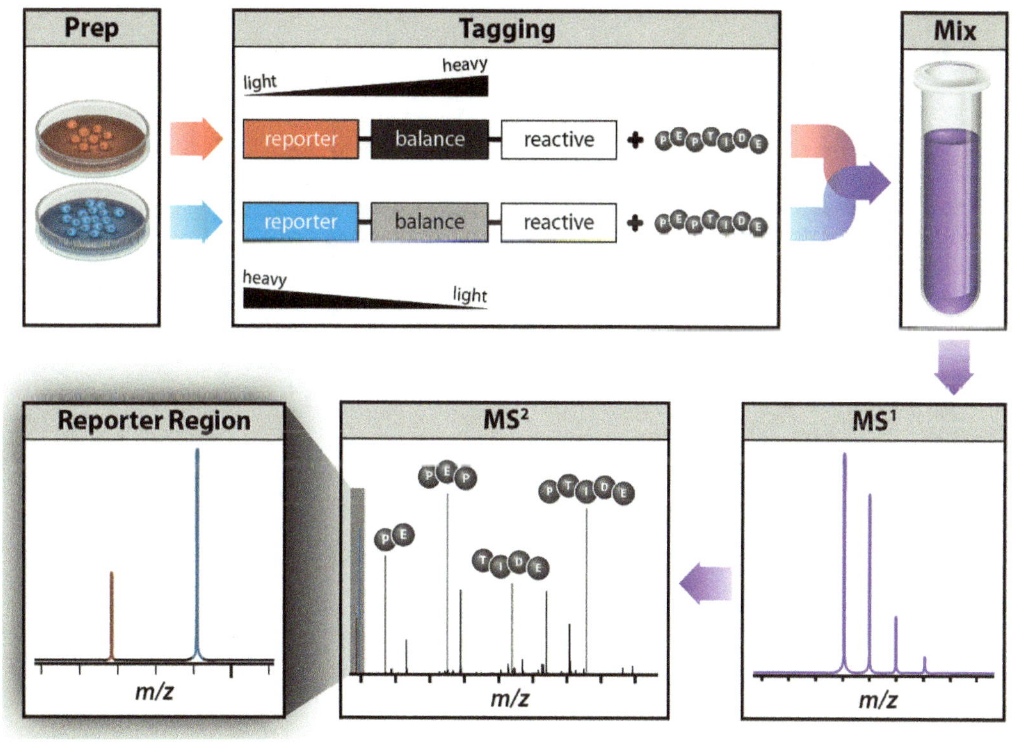

Isobaric labeling for tandem mass spectrometry: proteins are extracted from cells, digested, and labeled with tags of the same mass. When fragmented during MS/MS, the reporter ions show the relative amount of the peptides in the samples.

An isobaric tag for relative and absolute quantitation (iTRAQ) is a reagent for tandem mass spectrometry that is used to determine the amount of proteins from different sources in a single experiment. It uses stable isotope labeled molecules that can form a covalent bond with the N-terminus and side chain amines of proteins. The iTRAQ reagents are used to label peptides from different samples that are pooled and analyzed by liquid chromatography and tandem mass spectrometry. The fragmentation of the attached tag generates a low molecular mass reporter ion that can be used to relatively quantify the peptides and the proteins from which they originated.

Tandem Mass Tag

A tandem mass tag (TMT) is an isobaric mass tag chemical label used for protein quantification and identification. The tags contain four regions: mass reporter, cleavable linker, mass normalization, and protein reactive group.

Applications

Peptides

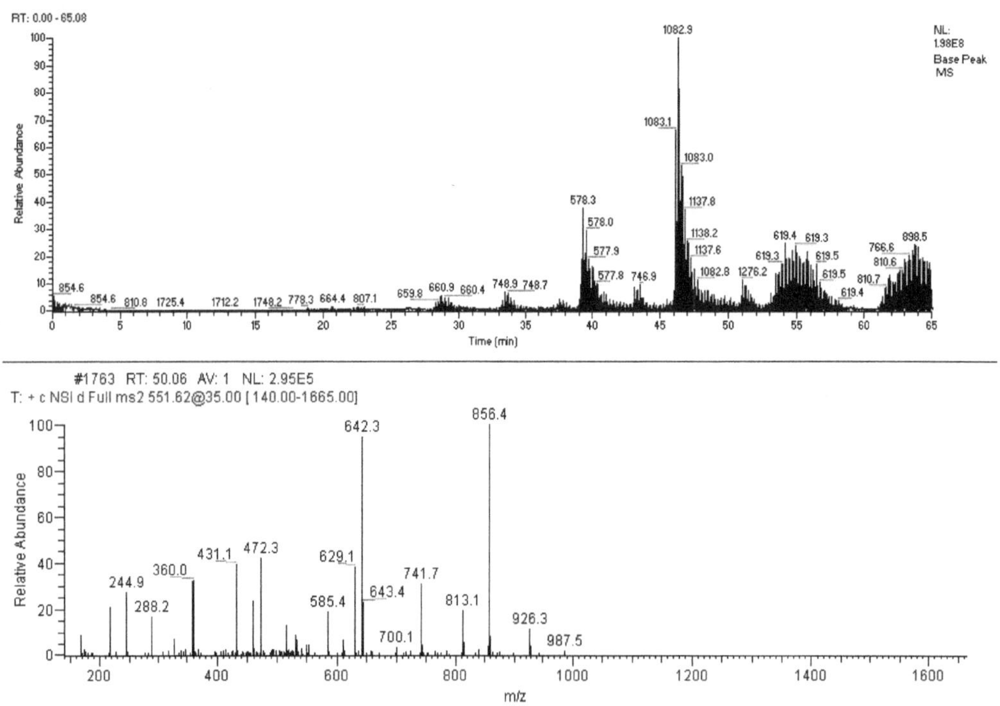

Chromatography trace (top) and tandem mass spectrum (bottom) of a peptide.

Tandem mass spectrometry can be used for protein sequencing. When intact proteins are introduced to a mass analyzer, this is called "top-down proteomics" and when proteins are digested into smaller peptides and subsequently introduced into the mass spectrometer, this is called "bottom-up proteomics". Shotgun proteomics is a variant of bottom up proteomics in which proteins in a mixture are digested prior to separation and tandem mass spectrometry.

Tandem mass spectrometry can produce a peptide sequence tag that can be used to identify a peptide in a protein database. A notation has been developed for indicating peptide fragments that arise from a tandem mass spectrum. Peptide fragment ions are indicated by a, b, or c if the charge is retained on the N-terminus and by x, y or z if the charge is maintained on the C-terminus. The subscript indicates the number of amino acid residues in the fragment. Superscripts are sometimes used to indicate neutral losses in addition to the backbone fragmentation, * for loss of ammonia and ° for loss of water. Although peptide backbone cleavage is the most useful for sequencing and peptide identification other fragment ions may be observed under high energy dissociation conditions. These include the side chain loss ions d, v, w and ammonium ions and additional sequence-specific fragment ions associated with particular amino acid residues.

Oligosaccharides

Oligosaccharides may be sequenced using tandem mass spectrometry in a similar manner to peptide sequencing. Fragmentation generally occurs on either side of the glycosidic bond (b, c, y and z ions) but also under more energetic conditions through the sugar ring structure in a cross-ring cleavage (x ions). Again trailing subscripts are used to indicate position of the cleavage along the chain. For cross ring cleavage ions the nature of the cross ring cleavage is indicated by preceding superscripts.

Oligonucleotides

Tandem mass spectrometry has been applied to DNA and RNA sequencing. A notation for gas-phase fragmentation of oligonucleotide ions has been proposed.

Newborn Screening

Newborn screening is the process of testing newborn babies for treatable genetic, endocrinologic, metabolic and hematologic diseases. The development of tandem mass spectrometry screening in the early 1990s led to a large expansion of potentially detectable congenital metabolic diseases that affect blood levels of organic acids.

ELECTROSTATIC PRECIPITATOR

Electrodes inside electrostatic precipitator.

Collection electrode of electrostatic precipitator in waste incineration plant.

An **electrostatic precipitator** (ESP) is a filtration device that removes fine particles, like dust and smoke, from a flowing gas using the force of an induced electrostatic charge minimally impeding the flow of gases through the unit.

In contrast to wet scrubbers which apply energy directly to the flowing fluid medium, an ESP applies energy only to the particulate matter being collected and therefore is very efficient in its consumption of energy (in the form of electricity).

Invention of the Electrostatic Precipitator

The first use of corona discharge to remove particles from an aerosol was by Hohlfeld in 1824. However, it was not commercialized until almost a century later.

In 1907 Frederick Gardner Cottrell, a professor of chemistry at the University of California, Berkeley, applied for a patent on a device for charging particles and then collecting them through electrostatic attraction—the first electrostatic precipitator. Cottrell first applied the device to the collection of sulphuric acid mist and lead oxide fumes emitted from various acid-making and smelting activities. Wine-producing vineyards in northern California were being adversely affected by the lead emissions.

At the time of Cottrell`s invention, the theoretical basis for operation was not understood. The operational theory was developed later in Germany, with the work of Walter Deutsch and the formation of the Lurgi company.

Cottrell used proceeds from his invention to fund scientific research through the creation of a foundation called Research Corporation in 1912, to which he assigned the patents. The intent of the organization was to bring inventions made by educators (such as Cottrell) into the commercial world for the benefit of society at large. The operation of Research Corporation is funded by royalties paid by commercial firms after commercialization occurs. Research Corporation has provided vital funding to many scientific projects: Goddard`s rocketry experiments, Lawrence`s cyclotron, production methods for vitamins A and B_1, among many others.

By a decision of the US Supreme Court. the Corporation had to be split into several entities. The Research Corporation was separated from two commercial firms making the hardware: Research-Cottrell Inc. (operating east of the Mississippi River) and Western Precipitation (operating in the western states). The Research Corporation continues to be active to this day, and the two companies formed to commercialize the invention for industrial and utility applications are still in business as well.

Electrophoresis is the term used for migration of gas-suspended charged particles in a direct-current electrostatic field. Traditional CRT television sets tend to accumulate dust on the screen because of this phenomenon (a CRT is a direct-current machine operating at about 35 kilovolts).

Plate Precipitator

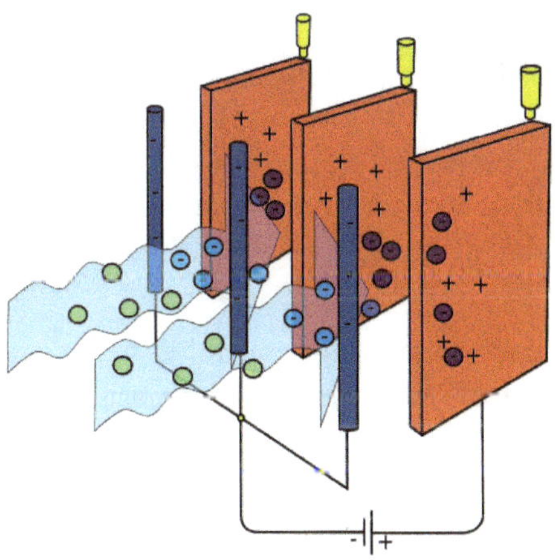

Conceptual diagram of an electrostatic precipitator.

The most basic precipitator contains a row of thin vertical wires, and followed by a stack of large flat metal plates oriented vertically, with the plates typically spaced about 1 cm to 18 cm apart, depending on the application. The air stream flows horizontally through the spaces between the wires, and then passes through the stack of plates.

A negative voltage of several thousand volts is applied between wire and plate. If the applied voltage is high enough, an electric corona discharge ionizes the air around the electrodes, which then ionizes the particles in the air stream.

The ionized particles, due to the electrostatic force, are diverted towards the grounded plates. Particles build up on the collection plates and are removed from the air stream.

A two-stage design (separate charging section ahead of collecting section) has the benefit of minimizing ozone production, which would adversely affect health of personnel working in enclosed spaces. For shipboard engine rooms where gearboxes generate an oil mist, two-stage ESP`s are used to clean the air, improving the operating environment and preventing buildup of flammable oil fog accumulations. Collected oil is returned to the gear lubricating system.

Collection Efficiency (*R*)

Precipitator performance is very sensitive to two particulate properties: 1) Electrical resistivity; and 2) Particle size distribution. These properties can be measured economically and accurately in the laboratory, using standard tests. Resistivity can be determined as a function of temperature in accordance with IEEE Standard 548. This test is conducted in an air environment containing a specified moisture concentration. The test is run as a function of ascending or descending temperature, or both. Data is acquired using an average ash layer[further explanation needed] electric field of 4 kV/cm. Since relatively low applied voltage is used and no sulfuric acid vapor is present in the test environment, the values obtained indicate the maximum ash resistivity.

In an ESP, where particle charging and discharging are key functions, resistivity is an important factor that significantly affects collection efficiency. While resistivity is an important phenomenon in the inter-electrode region where most particle charging takes place, it has a particularly important effect on the dust layer at the collection electrode where discharging occurs. Particles that exhibit high resistivity are difficult to charge. But once charged, they do not readily give up their acquired charge on arrival at the collection electrode. On the other hand, particles with low resistivity easily become charged and readily release their charge to the grounded collection plate. Both extremes in resistivity impede the efficient functioning of ESPs. ESPs work best under normal resistivity conditions.

Resistivity, which is a characteristic of particles in an electric field, is a measure of a particle's resistance to transferring charge (both accepting and giving up charges). Resistivity is a function of a particle's chemical composition as well as flue gas operating conditions such as temperature and moisture. Particles can have high, moderate (normal), or low resistivity.

Bulk resistivity is defined using a more general version of Ohm's Law, as given in Equation (**1**) below:

$$\vec{E} = \rho \vec{j}$$

Where:

E is the Electric field strength (V/cm);

j is the Current density (A/cm^2); and

ρ is the Resistivity (Ohm-cm)

A better way of displaying this would be to solve for resistivity as a function of applied voltage and current, as given in Equation (**2**) below:

$$\rho = \frac{AV}{Il}$$

Where:

ρ = Resistivity (Ohm-cm)

V = The applied DC potential, (Volts);

I = The measured current, (Amperes);

l = The ash layer thickness, (cm); and

A = The current measuring electrode face area, (cm^2).

Resistivity is the electrical resistance of a dust sample 1.0 cm^2 in cross-sectional area, 1.0 cm thick, and is recorded in units of ohm-cm. A method for measuring resistivity will be described in this article. The table below, gives value ranges for low, normal, and high resistivity.

Resistivity	Range of Measurement
Low	between 10^4 and 10^7 ohm-cm
Normal	between 10^7 and 2×10^{10} ohm-cm
High	above 2×10^{10} ohm-cm

Dust Layer Resistance

resistance affects electrical conditions in the dust layer by a potential electric field (voltage drop) being formed across the layer as negatively charged particles arrive at its surface and leak their electrical charges to the collection plate. At the metal surface of the electrically grounded collection plate, the voltage is zero, whereas at the outer surface of the dust layer, where new particles and ions are arriving, the electrostatic voltage caused by the gas ions can be quite high. The strength of this electric field depends on the resistance and thickness of the dust layer.

In high-resistance dust layers, the dust is not sufficiently conductive, so electrical charges have difficulty moving through the dust layer. Consequently, electrical charges accumulate on and beneath the dust layer surface, creating a strong electric field.

Voltages can be greater than 10,000 volts. Dust particles with high resistance are held too strongly to the plate, making them difficult to remove and causing rapping problems.

In low resistance dust layers, the corona current is readily passed to the grounded collection electrode. Therefore, a relatively weak electric field, of several thousand volts, is maintained across the dust layer. Collected dust particles with low resistance do not adhere strongly enough to the collection plate. They are easily dislodged and become retained in the gas stream.

The electrical conductivity of a bulk layer of particles depends on both surface and volume factors. Volume conduction, or the motions of electrical charges through the interiors of particles, depends mainly on the composition and temperature of the particles. In the higher temperature regions, above 500 °F (260 °C), volume conduction controls the conduction mechanism. Volume conduction also involves ancillary factors, such as compression of the particle layer, particle size and shape, and surface properties.

Volume conduction is represented in the figures as a straight-line at temperatures above 500 °F (260 °C). At temperatures below about 450 °F (230 °C), electrical charges begin to flow across surface moisture and chemical films adsorbed onto the particles. Surface conduction begins to lower the resistivity values and bend the curve downward at temperatures below 500 °F (260 °C).

These films usually differ both physically and chemically from the interiors of the particles owing to adsorption phenomena. Theoretical calculations indicate that moisture films only a few molecules thick are adequate to provide the desired surface conductivity. Surface conduction on particles is closely related to surface-leakage currents occurring on electrical insulators, which have been extensively studied. An interesting practical application of surface-leakage is the determination of dew point by measurement of the current between adjacent electrodes mounted on a glass surface. A sharp rise in current signals the formation of a moisture film on the glass. This method has been used effectively for determining the marked rise in dew point, which occurs when small amounts of sulfuric acid vapor are added to an atmosphere (commercial Dewpoint Meters are available on the market).

The following discussion of normal, high, and low resistance applies to ESPs operated in a dry state; resistance is not a problem in the operation of wet ESPs because of the moisture concentration in the ESP. The relationship between moisture content and resistance is explained later in this work.

Normal Resistivity

As stated above, ESPs work best under normal resistivity conditions. Particles with normal resistivity do not rapidly lose their charge on arrival at the collection electrode. These particles slowly leak their charge to grounded plates and are retained on the collection plates by intermolecular adhesive and cohesive forces. This allows a particulate layer to be built up and then dislodged from the plates by rapping. Within the range of normal dust resistivity (between 10^7 and 2 x 10^{10} ohm-cm), fly ash is collected more easily than dust having either low or high resistivity.

High Resistivity

If the voltage drop across the dust layer becomes too high, several adverse effects can occur. First, the high voltage drop reduces the voltage difference between the discharge electrode and collection electrode, and thereby reduces the electrostatic field strength used to drive the gas ion-charged particles over to the collected dust layer. As the dust layer builds up, and the electrical charges accumulate on the surface of the dust layer, the voltage difference between the discharge and collection electrodes decreases. The migration velocities of small particles are especially affected by the reduced electric field strength.

Another problem that occurs with high resistivity dust layers is called back corona. This occurs when the potential drop across the dust layer is so great that corona discharges begin to appear in the gas that is trapped within the dust layer. The dust layer breaks down electrically, producing small holes or craters from which back corona discharges occur. Positive gas ions are generated within the dust layer and are accelerated toward the "negatively charged" discharge electrode. The positive ions reduce some of the negative charges on the dust layer and neutralize some of the negative ions on the "charged particles" heading toward the collection electrode. Disruptions of the normal corona process greatly reduce the ESP's collection efficiency, which in severe cases, may fall below 50% . When back corona is present, the dust particles build up on the electrodes forming a layer of insulation. Often this can not be repaired without bringing the unit offline.

The third, and generally most common problem with high resistivity dust is increased electrical sparking. When the sparking rate exceeds the "set spark rate limit," the automatic controllers limit the operating voltage of the field. This causes reduced particle charging and reduced migration velocities toward the collection electrode. High resistivity can generally be reduced by doing the following:

- Adjusting the temperature;
- Increasing moisture content;
- Adding conditioning agents to the gas stream;
- Increasing the collection surface area; and
- Using hot-side precipitators (occasionally and with foreknowledge of sodium depletion).

Thin dust layers and high-resistivity dust especially favor the formation of back corona craters. Severe back corona has been observed with dust layers as thin as 0.1 mm, but a dust layer just over one particle thick can reduce the sparking voltage by 50%. The most marked effects of back corona on the current-voltage characteristics are:

- Reduction of the spark over voltage by as much as 50% or more;
- Current jumps or discontinuities caused by the formation of stable back-corona craters; and
- Large increase in maximum corona current, which just below spark over corona gap may be several times the normal current.

The Figure below and to the left shows the variation in resistivity with changing gas temperature for six different industrial dusts along with three coal-fired fly ashes. The Figure on the right illustrates resistivity values measured for various chemical compounds that were prepared in the laboratory.

Results for Fly Ash A (in the figure to the left) were acquired in the ascending temperature mode. These data are typical for a moderate to high combustibles content ash. Data for Fly Ash B are from the same sample, acquired during the descending temperature mode.

The differences between the ascending and descending temperature modes are due to the presence of unburned combustibles in the sample. Between the two test modes, the samples are equilibrated in dry air for 14 hours (overnight) at 850 °F (450 °C). This overnight annealing process typically removes between 60% and 90% of any unburned combustibles present in the samples. Exactly how carbon works as a charge carrier is not fully understood, but it is known to significantly reduce the resistivity of a dust.

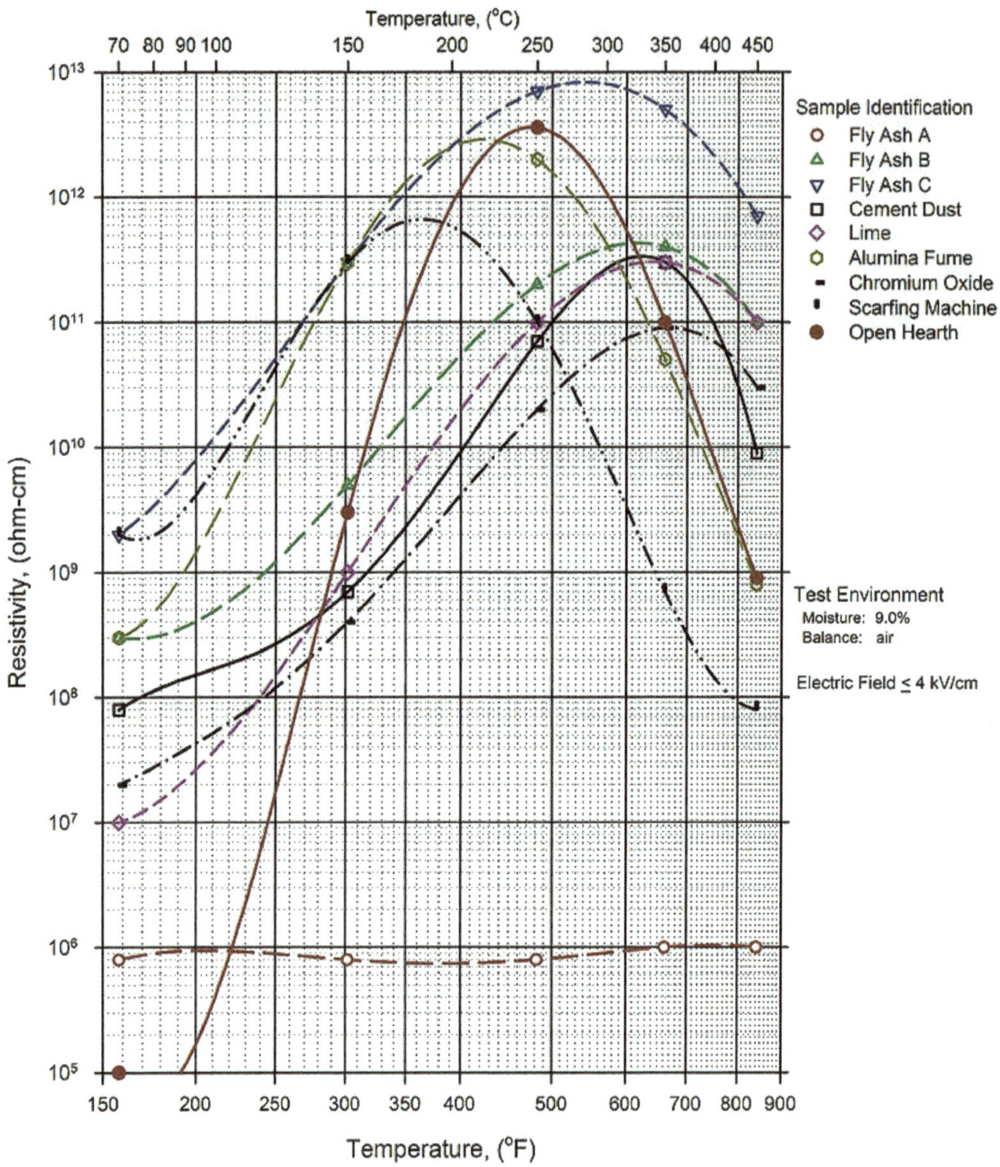

Resistivity Values of Representative Dusts and Fumes From Industrial Plants.

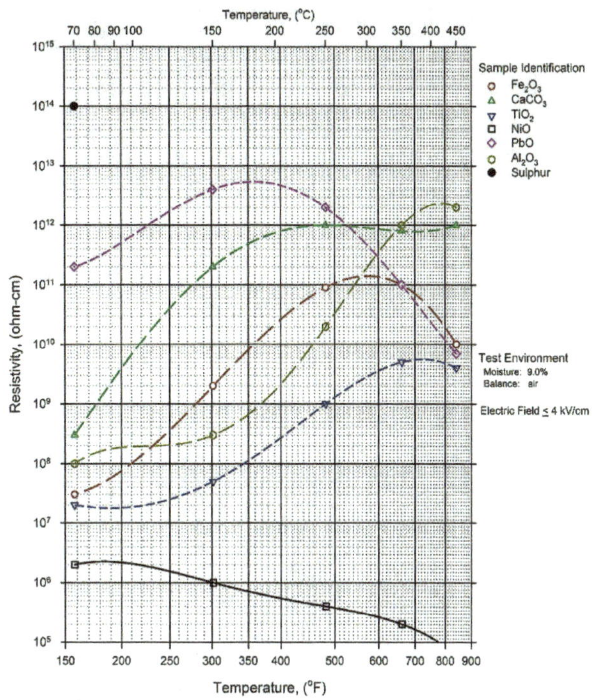

Resistivity Values of Various Chemicals and Reagents as a Function of Temperature.

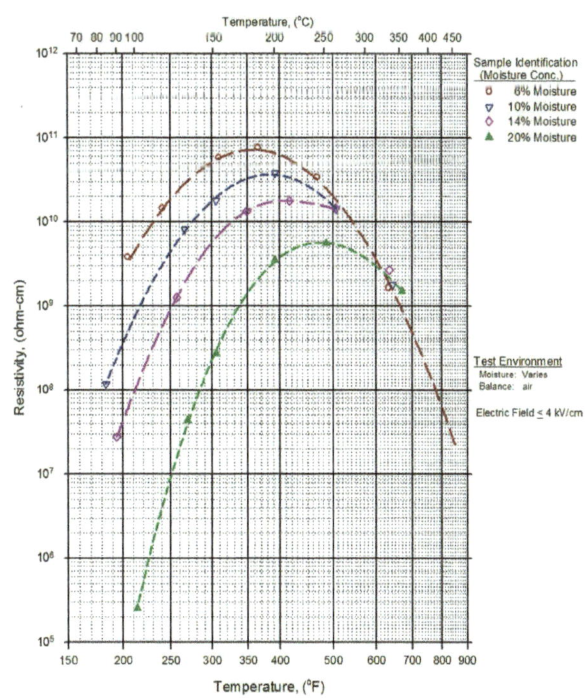

Resistivity Measured as a Function of Temperature in Varying Moisture Concentrations (Humidity).

Carbon can act, at first, like a high resistivity dust in the precipitator. Higher voltages can be required in order for corona generation to begin. These higher voltages can be problematic for the TR-Set controls. The problem lies in onset of corona causing large amounts of current to surge through the (low resistivity) dust layer. The controls sense this surge as a spark. As precipitators are operated in spark-limiting mode, power is terminated and the corona generation cycle re-initiates. Thus, lower power (current) readings are noted with relatively high voltage readings.

The same thing is believed to occur in laboratory measurements. Parallel plate geometry is used in laboratory measurements without corona generation. A stainless steel cup holds the sample. Another stainless steel electrode weight sits on top of the sample (direct contact with the dust layer). As voltage is increased from small amounts (e.g. 20 V), no current is measured. Then, a threshold voltage level is reached. At this level, current surges through the sample... so much so that the voltage supply unit can trip off. After removal of the unburned combustibles during the above-mentioned annealing procedure, the descending temperature mode curve shows the typical inverted “V” shape one might expect.

Low Resistivity

Particles that have low resistivity are difficult to collect because they are easily charged (very conductive) and rapidly lose their charge on arrival at the collection electrode. The particles take on the charge of the collection electrode, bounce off the plates, and become re-entrained in the gas stream. Thus, attractive and repulsive electrical forces that are normally at work at normal and higher resistivities are lacking, and the binding forces to the plate are considerably lessened. Examples of low-resistivity dusts are unburned carbon in fly ash and carbon black.

If these conductive particles are coarse, they can be removed upstream of the precipitator by using a device such as a cyclone mechanical collector.

The addition of liquid ammonia (NH_3) into the gas stream as a conditioning agent has found wide use in recent years. It is theorized that ammonia reacts with H_2SO_4 contained in the flue gas to form an ammonium sulfate compound that increases the cohesivity of the dust. This additional cohesivity makes up for the loss of electrical attraction forces.

The table below summarizes the characteristics associated with low, normal and high resistivity dusts.

The moisture content of the flue gas stream also affects particle resistivity. Increasing the moisture content of the gas stream by spraying water or injecting steam into the duct work preceding the ESP lowers the resistivity. In both temperature adjustment and moisture conditioning, one must maintain gas conditions above the dew point to prevent corrosion problems in the ESP or downstream equipment. The figure to the right shows the effect of temperature and moisture on the resistivity of a cement dust. As the percentage of moisture in the gas stream increases from 6 to 20%, the resistivity of the dust dramatically decreases. Also, raising or lowering the temperature can decrease cement dust resistivity for all the moisture percentages represented.

The presence of SO_3 in the gas stream has been shown to favor the electrostatic precipitation process when problems with high resistivity occur. Most of the sulfur content in the coal burned for combustion sources converts to SO_2. However, approximately 1% of the sulfur converts to SO_3. The amount of SO_3 in the flue gas normally increases with increasing sulfur content of the coal. The resistivity of the particles decreases as the sulfur content of the coal increases.

Resistivity	Range of Measurement	Precipitator Characteristics
Low	between 10^4 and 10^7ohm-cm	1. Normal operating voltage and current levels unless dust layer is thick enough to reduce plate clearances and cause higher current levels. 2. Reduced electrical force component retaining collected dust, vulnerable to high reentrainment losses. 3. Negligible voltage drop across dust layer. 4. Reduced collection performance due to (2)
Normal	between 10^7 and 2×10^{10} ohm-cm	1. Normal operating voltage and current levels. 2. Negligible voltage drop across dust layer. 3. Sufficient electrical force component retaining collected dust. 4. High collection performance due to (1), (2) and (3)
Marginal to High	between $2 \times 10^{10}$and 10^{12} ohm-cm	1. Reduced operating voltage and current levels with high spark rates. 2. Significant voltage loss across dust layer. 3. Moderate electrical force component retaining collected dust. 4. Reduced collection performance due to (1) and (2)
High	above 10^{12} ohm-cm	1. Reduced operating voltage levels; high operating current levels if power supply controller is not operating properly. 2. Very significant voltage loss across dust layer. 3. High electrical force component retaining collected dust. 4. Seriously reduced collection performance due to (1), (2) and probably back corona.

Other conditioning agents, such as sulfuric acid, ammonia, sodium chloride, and soda ash (sometimes as raw trona), have also been used to reduce particle resistivity. Therefore, the chemical composition of the flue gas stream is important with regard to the resistivity of the particles to be collected in the ESP. The table below lists various conditioning agents and their mechanisms of operation.

Conditioning Agent	Mechanism(s) of Action
Sulfur Trioxide and/or Sulfuric Acid	1. Condensation and adsorption on fly ash surfaces. 2. may also increase cohesiveness of fly ash. 3. Reduces resistivity
Ammonia	Mechanism is not clear, various ones proposed; 1. Modifies resistivity. 2. Increases ash cohesiveness. 3. Enhances space charge effect.
Ammonium Sulfate	Little is known about the mechanism; claims are made for the following: 1. Modifies resistivity (depends upon injection temperature). 2. Increases ash cohesiveness. 3. Enhances space charge effect. 4. Experimental data lacking to substantiate which of these is predominant.
Triethylamine	Particle agglomeration claimed; no supporting data.
Sodium Compounds	1. Natural conditioner if added with coal. 2. Resistivity modifier if injected into gas stream.
Compounds of Transition Metals	Postulated that they catalyze oxidation of SO_2 to SO_3; no definitive tests with fly ash to verify this postulation.
Potassium Sulfate and Sodium Chloride	In cement and lime kiln ESPs: 1. Resistivity modifiers in the gas stream. 2. NaCl - natural conditioner when mixed with coal.

If injection of ammonium sulfate occurs at a temperature greater than about 600 °F (320 °C), dissociation into ammonia and sulfur trioxide results. Depending on the ash, SO2 may preferentially interact with fly ash as SO3 conditioning. The remainder recombines with ammonia to add to the space charge as well as increase cohesiveness of the ash.

More recently, it has been recognized that a major reason for loss of efficiency of the electrostatic precipitator is due to particle buildup on the charging wires in addition to the collection plates (Davidson and McKinney, 1998). This is easily remedied by making sure that the wires themselves are cleaned at the same time that the collecting plates are cleaned.

Sulfuric acid vapor (SO_3) enhances the effects of water vapor on surface conduction. It is physically adsorbed within the layer of moisture on the particle surfaces. The effects of relatively small amounts of acid vapor can be seen in the figure below and to the right.

The inherent resistivity of the sample at 300 °F (150 °C) is 5×10^{12} ohm-cm. An equilibrium concentration of just 1.9 ppm sulfuric acid vapor lowers that value to about 7×10^9 ohm-cm.

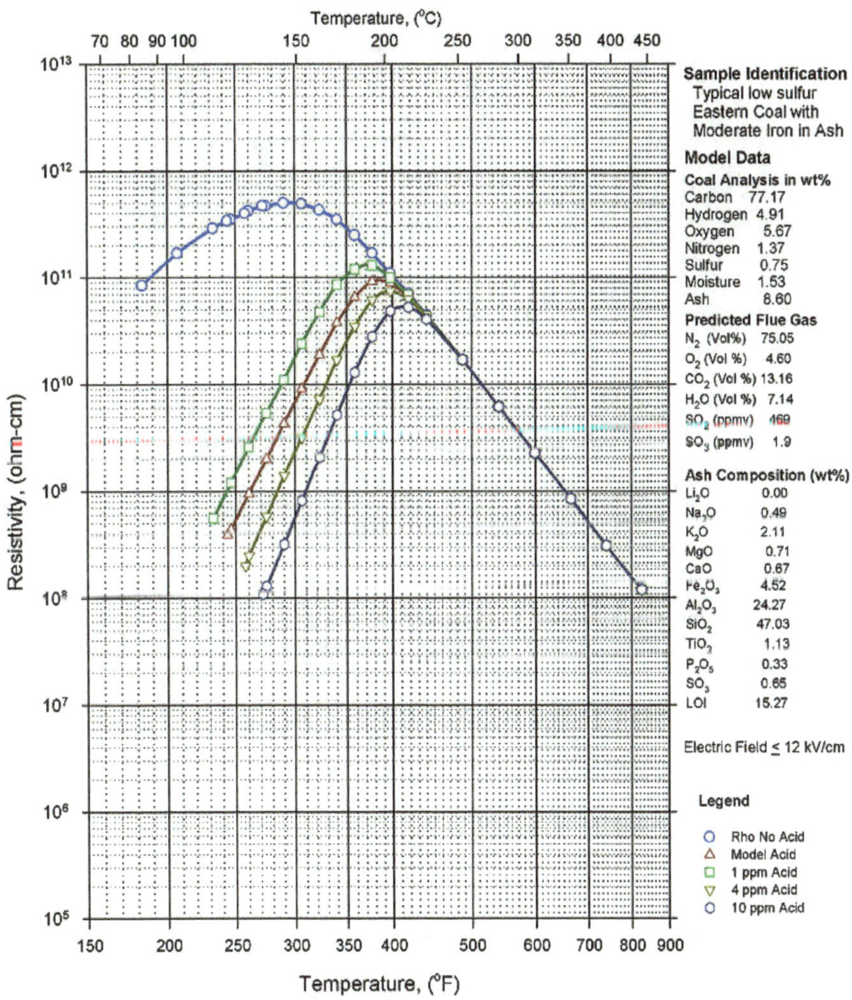

Resistivity Modeled As A Function of Environmental Conditions - Especially Sulfuric Acid Vapor

Modern Industrial Electrostatic Precipitators

A smokestack at coal-fired Hazelwood Power Station in Victoria, Australia emits brown smoke when its ESP is shut down.

ESPs continue to be excellent devices for control of many industrial particulate emissions, including smoke from electricity-generating utilities (coal and oil fired), salt cake collection from black liquor boilers in pulp mills, and catalyst collection from fluidized bed catalytic cracker units in oil refineries to name a few. These devices treat gas volumes from several hundred thousand ACFM to 2.5 million ACFM (1,180 m^3/s) in the largest coal-fired boiler applications. For a coal-fired boiler the collection is usually performed downstream of the air preheater at about 160 °C (320 °F) which provides optimal resistivity of the coal-ash particles. For some difficult applications with low-sulfur fuel hot-end units have been built operating above 370 °C (698 °F).

The original parallel plate–weighted wire design has evolved as more efficient (and robust) discharge electrode designs were developed, today focusing on rigid (pipe-frame) discharge electrodes to which many sharpened spikes are attached (barbed wire), maximizing corona production. Transformer-rectifier systems apply voltages of 50–100 kV at relatively high current densities. Modern controls, such as an automatic voltage control, minimize electric sparking and prevent arcing (sparks are quenched within 1/2 cycle of the TR set), avoiding damage to the components. Automatic plate-rapping systems and hopper-evacuation systems remove the collected particulate matter while on line, theoretically allowing ESPs to stay in continuous operation for years at a time.

Electrostatic Air Sampling

Electrostatic precipitators can be used to sample airborne particles or aerosol for analysis purposes. Precipitator designs with a liquid counterelectrode can be used to sample biological particles, e.g. viruses, directly into a small liquid volume to reduce unnecessary sample dilution.

Wet Electrostatic Precipitator

A wet electrostatic precipitator (WESP or wet ESP) operates with water vapor saturated air streams (100% relative humidity). WESPs are commonly used to remove liquid droplets such as sulfuric acid mist from industrial process gas streams. The WESP is also commonly used where the gases are high in moisture content, contain combustible particulate, or have particles that are sticky in nature.

The preferred and most modern type of WESP is a downflow tubular design. This design allows the collected moisture and particulate to form a moving slurry that helps to keep the collection surfaces clean. Plate style and upflow design WESPs are very unreliable and should not be used in applications where particulate is sticky in nature.

Consumer-Oriented Electrostatic Air Cleaners

A portable electrostatic air cleaner marketed to consumers

Portable electrostatic air cleaner with cover removed, showing collector plates.

Plate precipitators are commonly marketed to the public as air purifier devices or as a permanent replacement for furnace filters, but all have the undesirable attribute of being somewhat messy to clean. A negative side-effect of electrostatic precipitation devices is the potential production of toxic ozone and NO_x. However, electrostatic precipitators offer benefits over other air purifications technologies, such as HEPA filtration, which require expensive filters and can become "production sinks" for many harmful forms of bacteria.

With electrostatic precipitators, if the collection plates are allowed to accumulate large amounts of particulate matter, the particles can sometimes bond so tightly to the metal plates that vigorous washing and scrubbing may be required to completely clean the collection plates. The close spacing of the plates can make thorough cleaning difficult, and the stack of plates often cannot be easily disassembled for cleaning. One solution, suggested by several manufacturers, is to wash the collector plates in a dishwasher.

Some consumer precipitation filters are sold with special soak-off cleaners, where the entire plate array is removed from the precipitator and soaked in a large container overnight, to help loosen the tightly bonded particulates.

A study by the Canada Mortgage and Housing Corporation testing a variety of forced-air furnace filters found that ESP filters provided the best, and most cost-effective means of cleaning air using a forced-air system.

The first portable electrostatic air filter systems for homes was marketed in 1954 by Raytheon.

MICROFILTRATION

Microfiltration (commonly abbreviated to MF) is a type of physical filtration process where a contaminated fluid is passed through a special pore-sized membrane to separate microorganisms and suspended particles from process liquid. It is commonly used in conjunction with various other separation processes such as ultrafiltration and reverse osmosis to provide a product stream which is free of undesired contaminants.

General Principles

Microfiltration usually serves as a pre-treatment for other separation processes such as ultrafiltration, and a post-treatment for granular media filtration. The typical particle size used for microfiltration ranges from about 0.1 to 10 µm. In terms of approximate molecular weight these membranes can separate macromolecules of molecular weights generally less than 100,000 g/mol. The filters used in the microfiltration process are specially designed to prevent particles such as, sediment, algae, protozoa or large bacteria from passing through a specially designed filter. More microscopic, atomic or ionic materials such as water (H_2O), monovalent species such as Sodium (Na^+) or Chloride (Cl^-) ions, dissolved or natural organic matter, and small colloids and viruses will still be able to pass through the filter.

The suspended liquid is passed though at a relatively high velocity of around 1–3 m/s and at low to moderate pressures (around 100-400 kPa) parallel or tangential to the semi-permeable membrane in a sheet or tubular form. A pump is commonly fitted onto the processing equipment to allow the liquid to pass through the membrane filter. There are also two pump configurations, either pressure driven or vacuum. A differential or regular pressure gauge is commonly attached to measure the pressure drop between the outlet and inlet streams. See Figure 1 for a general setup.

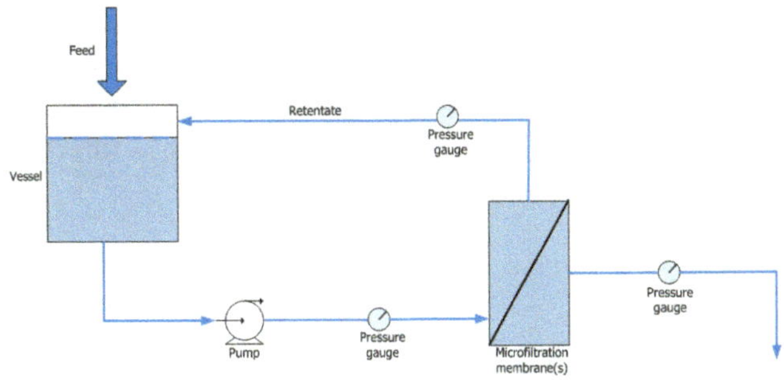

Figure 1: Overall setup for a microfiltration system.

The most abundant use of microfiltration membranes are in the water, beverage and bio-processing industries (see below). The exit process stream after treatment using a micro-filter has a recovery rate which generally ranges to about 90-98 %.

Range of Applications

Water Treatment

Perhaps the most prominent use of microfiltration membranes pertains to the treatment of potable water supplies. The membranes are a key step in the primary disinfection of the uptake water stream. Such a stream might contain pathogens such as the protozoa Cryptosporidium and Giardia lamblia which are responsible for numerous disease outbreaks. Both species show a gradual resistance to traditional disinfectants (i.e. chlorine). The use of MF membranes presents a physical means of separation (a barrier) as opposed to a chemical alternative. In this sense, both filtration and disinfection take place in a single step, negating the extra cost of chemical dosage and the corresponding equipment (needed for handling and storage).

Similarly, the MF membranes are used in secondary wastewater effluents to remove turbidity but also to provide treatment for disinfection. At this stage, coagulants (iron or aluminum) may potentially be added to precipitate species such as phosphorus and arsenic which would otherwise have been soluble.

Sterilization

Another crucial application of MF membranes lies in the cold sterilisation of beverages and pharmaceuticals. Historically, heat was used to sterilize refreshments such as juice, wine and beer in particular, however a palatable loss in flavour was clearly evident upon heating. Similarly, pharmaceuticals have been shown to lose their effectiveness upon heat addition. MF membranes are employed in these industries as a method to remove bacteria and other undesired suspensions from liquids, a procedure termed as 'cold sterilisation', which negate the use of heat.

Petroleum Refining

Furthermore, microfiltration membranes are finding increasing use in areas such as petroleum refining, in which the removal of particulates from flue gases is of particular concern. The key challenges/requirements for this technology are the ability of the membrane modules to withstand high temperatures (i.e. maintain stability), but also the design must be such to provide a very thin sheeting (thickness < 2000 angstroms) to facilitate an increase of flux. In addition the modules must have a low fouling profile and most importantly, be available at a low-cost for the system to be financially viable.

Dairy Processing

Aside from the above applications, MF membranes have found dynamic use in major areas within the dairy industry, particularly for milk and whey processing. The MF membranes aid in the removal of bacteria and the associated spores from milk, by rejecting the harmful species from passing through. This is also a precursor for pasteurisation, allowing for an extended shelf-life of the product. However, the most promising technique for MF membranes in this field pertains to the separation of casein from whey proteins (i.e. serum milk proteins). This results in two product streams both of which are highly relied on by consumers; a casein-rich concentrate stream used for cheese making, and a whey/serum protein stream which is further processed (using ultrafiltration) to make whey protein concentrate. The whey protein stream undergoes further filtration to remove fat in order to achieve higher protein content in the final WPC (Whey Protein Concentrate) and WPI (Whey Protein Isolate) powders.

Other Applications

Other common applications utilising microfiltration as a major separation process include

- Clarification and purification of cell broths where macromolecules are to be separated from other large molecules, proteins, or cell debris.
- Other biochemical and bio-processing applications such as clarification of dextrose.
- Production of Paints and Adhesives.

Main Process Characteristics

Membrane filtration processes can be distinguished by three major characteristics; Driving force, retentate stream and permeate streams. The microfiltration process is pressure driven with suspended particles and water as retentate and dissolved solutes plus water as permeate. The use of hydraulic pressure accelerates the separation process by increasing the flow rate (flux) of the liquid stream but does not affect the chemical composition of the species in the retentate and product streams.

A major characteristic that limits the performance of microfiltration or any membrane technology is a process known as fouling. Fouling describes the deposition and accumulation of feed components such as suspended particles, impermeable dissolved solutes or even permeable solutes, on the membrane surface and or within the pores of the membrane. Fouling of the membrane during the filtration processes decreases the flux and thus overall efficiency of the operation. This is indicated when the pressure drop increases to a certain point. It occurs even when operating parameters are constant (pressure, flow rate, temperature and concentration) Fouling is mostly irreversible although a portion of the fouling layer can be reversed by cleaning for short periods of time.

Microfiltration membranes can generally operate in one of two configurations.

Cross-flow filtration: where the fluid is passed through tangentially with respect to the membrane. Part of the feed stream containing the treated liquid is collected below the filter while parts of the water are passed through the membrane untreated. Cross flow filtration is understood to be a unit operation rather than a process. *Refer to Figure 2 for a general schematic for the process.*

Dead-end filtration; all of the process fluid flows and all particles larger than the pore sizes of the membrane are stopped at its surface. All of the feed water is treated at once subject to cake formation. This process is mostly used for batch or semicontinuous filtration of low concentrated solutions, *Refer to Figure 3 for a general schematic for this process.*

Process and Equipment Design

The major issues which influence the selection of the membrane include

Site Specific Issues (Unique to the site where the plant is located)

- Capacity and demand of the facility.
- Percentage recovery and rejection.
- Fluid characteristics (viscosity, turbidity, density)
- Quality of the fluid to be treated
- Pre-treatment processes

Membrane Specific Issues (Unique to the manufacturer or supplier)

- Cost of material procurement and manufacture
- Operating temperature
- Trans-membrane pressure
- Membrane flux
- Handling fluid characteristics (Viscosity, Turbidity, Density)
- Monitoring and maintenance of the system
- Cleaning and treatment
- Disposal of process residuals

Process Design Variables (Regarding proper membrane selection)

- Operation and control of all processes in the system,
- Materials of construction
- Equipment and instrumentation (controllers, sensors) and their cost.

Fundamental Design Heuristics

A few important design heuristics and their assessment are discussed below:

- When treating raw contaminated fluids, hard sharp materials can wear and tear the porous cavities in the micro-filter, rendering it ineffective. Liquids must be subjected to pre-treatment before passage through the micro-filter. This may be achieved by a variation of macro separation processes such as screening, or granular media filtration.
- When undertaking cleaning regimes the membrane must not dry out once it has been contacted by the process stream.Thorough water rinsing of the membrane modules, pipelines, pumps and other unit connections should be carried out until the end water appears clean.
- Microfiltration modules are typically set to operate at pressures of 100 to 400 kPa. These pressures allow removal of materials such as sand, slits and clays, and also bacteria and protozoa.
- When the membrane modules are being used for the first time, i.e. during plant start-up, conditions need to be well devised. Generally a slow-start is required when the feed is introduced into the modules, since even slight perturbations above the critical flux will result in irreversible fouling.

Like any other membranes Microfiltration membranes are prone to fouling. *(See Figure 4 below)* It is therefore necessary that regular maintenance be carried out to prolong the life of the membrane module.

- Routine 'backwashing', is used to achieve this. Depending on the specific application of the membrane, backwashing is carried out in short durations (typically 3 to 180 s) and in moderately frequent intervals (5 min to several hours). Turbulent flow conditions with Reynolds numbers greater than 2100, ideally between 3000 - 5000 should be used. This should not however be confused with 'backflushing', a more rigorous and thorough cleaning technique, commonly practiced in cases of particulate and colloidal fouling.
- When major cleaning is needed to remove entrained particles, a CIP (Clean In Place) technique is used. Cleaning agents/detergents, such as sodium hypochlorite, citric acid, caustic soda or even special enzymes are typically used for this purpose. The concentration of these chemicals is dependent on the type of the membrane (its sensitivity to strong chemicals), but also the type of matter (e.g. scaling due to the presence of calcium ions) to be removed.
- Another method to increase the lifespan of the membrane may be feasible to design two microfiltration membranes in series. The first filter would be used for pre-treatment of the liquid passing though the membrane, where larger particles and deposits are captured on the cartridge. The second filter would act as an extra "check" for particles which are able to pass through the first membrane as well as provide screening for particles on the lower spectrum of the range.

Design Economics

The cost to design and manufacture a membrane per unit of area are about 20% less compared to the early 1990s and in a general sense are constantly declining. Microfiltration membranes are more advantageous in comparison to conventional systems. Microfiltration systems do not require expensive extraneous equipment such as flocculates, addition of chemicals, flash mixers, settling and filter basins. However the cost of replacement of capital equipment costs (membrane cartridge filters etc.) might still be relatively high as the equipment may be manufactured specific to the application. Using the design heuristics and general plant design principles (mentioned above), the membrane life-span can be increased to reduce these costs.

Through the design of more intelligent process control systems and efficient plant designs some general tips to reduce operating costs are listed below

- Running plants at reduced fluxes or pressures at low load periods (winter)
- Taking plant systems off-line for short periods when the feed conditions are extreme.
- A short shutdown period (approximately 1 hour) during the first flush of a river after rainfall (in water treatment applications) to reduce cleaning costs in the initial period.
- The use of more cost effective cleaning chemicals where suitable (sulphuric acid instead of citric/ phosphoric acids.)
- The use of a flexible control design system. Operators are able to manipulate variables and setpoints to achieve maximum cost savings.

Table 1 (below) expresses an indicative guide of membrane filtration capital and operating costs per unit of flow.

Table 1 Approximate Costing of Membrane Filtration per unit of flow

Parameter	Amount	Amount	Amount	Amount	Amount
Design Flow (mg/d)	0.01	0.1	1.0	10	100
Average Flow (mg/d)	0.005	0.03	0.35	4.4	50
Capital Cost ($/gal)	$18.00	$4.30	$1.60	$1.10	$0.85
Annual Operating and Managing Costs ($/kgal)	$4.25	$1.10	$0.60	$0.30	$0.25

Note:

- *Capital Costs are based on dollars per gallon of the treatment plant capacity*
- Design flow is measured in millions of gallons per day.
- Membrane Costs only (No Pre-Treatment or Post-Treatment equipment considered in this table)
- Operating and Annual costs, are based on dollars per thousand gallons treated.
- All prices are in US dollars current of 2009, and is not adjusted for inflation.

Process Equipment

Membrane Materials

The materials which constitute the membranes used in microfiltration systems may be either organic or inorganic depending upon the contaminants that are desired to be removed, or the type of application.

- Organic membranes are made using a diverse range of polymers including cellulose acetate (CA), polysulfone, polyvinylidene fluoride, polyethersulfone and polyamide. These are most commonly used due to their flexibility, and chemical properties.
- Inorganic membranes are usually composed of sintered metal or porous alumina. They are able to be designed in various shapes, with a range of average pore sizes and permeability.

Membrane Equipment

General Membrane structures for microfiltration include

- Screen filters (Particles and matter which are of the same size or larger than the screen openings are retained by the process and are collected on the surface of the screen)
- Depth filters (Matter and particles are embedded within the constrictions within the filter media, the filter surface contains larger particles, smaller particles are captured in a narrower and deeper section of the filter media.)

Microfiltration Membrane Modules

Plate and frame (flat sheet)

Membrane modules for dead-end flow microfiltration are mainly plate-and-frame configurations. They possess a flat and thin-film composite sheet where the plate is asymmetric. A thin selective skin is supported on a thicker layer that has larger pores. These systems are compact and possess a sturdy design, Compared to cross-flow filtration, plate and frame configurations possess a reduced capital expenditure; however the operating costs will be higher. The uses of plate and frame modules are most applicable for smaller and simpler scale applications (laboratory) which filter dilute solutions.

Spiral-wound

This particular design is used for cross-flow filtration. The design involves a pleated membrane which is folded around a perforated permeate core, akin to a spiral, that is usually placed within a pressure vessel. This particular design is preferred when the solutions handled is heavily concentrated and in conditions of high temperatures and extreme pH. This particular configuration is generally used in more large scale industrial applications of microfiltration.

Fundamental Design Equations

As separation is achieved by sieving, the principal mechanism of transfer for microfiltration through micro porous membranes is bulk flow.

Generally, due to the small diameter of the pores the flow within the process is laminar (Reynolds Number < 2100) The flow velocity of the fluid moving through the pores can thus be determined (by Hagen-Poiseuille's equation), the simplest of which assuming a parabolic velocity profile.

$$v = \frac{D^2 * \Delta P}{32 * \mu * L}$$

Transmembrane Pressure (TMP)

The transmembrane pressure (TMP) is defined as the mean of the applied pressure from the feed to the concentrate side of the membrane subtracted by the pressure of the permeate. This is applied to dead-end filtration mainly and is indicative of whether a system is fouled sufficiently to warrant replacement.

$$v = \frac{P_F + P_C}{2} - P_P$$

Where

- P_f is the pressure on the Feed Side
- P_c is the pressure of the Concentrate
- P_p is the pressure of the Permeate

Permeate Flux

The permeate flux in microfiltration is given by the following relation, based on Darcy's Law

$$J_v = \frac{1}{A_M} * \frac{dV}{dt} = \frac{\Delta P}{\mu * (R_u + R_c)}$$

Where

- R_u= Permeate membrane flow resistance $(m-1)$
- R_c= Permeate cake resistance $(m-1)$
- μ = Permeate viscosity (kg m-1 s-1)
- ΔP = Pressure Drop between the cake and membrane

The cake resistance is given by:

$$R_c = r * \frac{V_S}{A_m}$$

Where

- r = Specific cake resistance (m-2)
- Vs = Volume of cake (m3)
- AM = Area of membrane (m2)

For micron sized particles the Specific Cake Resistance is roughly.

$$r = \frac{180 * (1 - \epsilon)}{\epsilon^3 * d_s^2}$$

Where

- ε = Porosity of cake (unitless)
- d_s = Mean particle diameter (m)

Rigorous Design Equations

To give a better indication regarding the exact determination of the extent of the cake formation, one-dimensional quantitative models have been formulated to determine factors such as

- Complete Blocking (Pores with an initial radius less that the radius of the pore)
- Standard Blocking
- Sublayer Formation
- Cake Formation

See External Links for further details

Environmental Issues, Safety and Regulation

Although environmental impacts of membrane filtration processes differs between applications a generic method of evaluation is the Life-cycle assessment (LCA), a tool for the analysis of the environmental burden of membrane filtration processes at all stages. It accounts for all types of impacts upon the environment including emission to land, water and air.

In regards to microfiltration processes there are a number of potential environmental impacts to be considered. They include: global warming potential, photo-oxidant formation potential, eutrophication potential, human toxicity potential, freshwater ecotoxicity potential, marine ecotoxicity potential and terrestrial ecotoxicity potential. In general, the potential environmental impact of the process is largely dependent on flux and the maximum transmembrane pressure, however other operating parameters remain a factor to be considered. A specific comment on which exact combination of operational condition will yield the lowest burden on the environment cannot be made as each application will require different optimisations.

In a general sense, membrane filtration processes are relative “low risk” operations, that is, the potential for dangerous hazards are small. There are, however several aspects to be mindful of. All pressure-driven filtration processes including microfiltration requires a degree of pressure to be applied to the feed liquid stream as well as imposed electrical concerns. Other factors contributing to safety are dependent on parameters of the process. For example, processing dairy product will lead to bacteria formations that must be controlled to comply with safety and regulatory standards.

Comparison with Similar Processes

Membrane microfiltration is fundamentally the same as other filtration techniques utilising a pore size distribution to physically separate particles. It is analogous to other technologies such as ultra/nanofiltration and reverse osmosis, however, the only difference exists in the size of the particles retained, and also the osmotic pressure. The main of which are described in general below:

Ultrafiltration (UF)

Ultrafiltration membranes have pore sizes ranging from 0.1 μm to 0.01 μm and are able to retain proteins, endotoxins, viruses and silica. UF has diverse applications which span from waste water treatment to pharmaceutical applications.

Nanofiltration (NF)

Nanofiltration membranes have pores sized from 0.001 μm to 0.01 μm and filters multivalent ions, synthetic dyes, sugars and specific salts. As the pore size drops from MF to NF, the osmotic pressure requirement increases.

Reverse Osmosis (RO)

Reverse Osmosis is the finest separation membrane process available, pore sizes range from 0.0001 μm to 0.001 μm. RO is able to retain mostly all molecules except for water and due to the size of the pores, the required osmotic pressure is significantly greater than that for MF. Both reverse osmosis and nanofiltration are fundamentally different since the flow goes against the concentration gradient, because those systems use pressure as a means of forcing water to go from low pressure to high pressure.

Recent Developments

Recent advances in MF have focused on manufacturing processes for the construction of membranes and additives to promote coagulation and therefore reduce the fouling of the membrane. Since MF, UF, NF and RO are closely related, these advances are applicable to multiple processes and not MF alone.

Recently studies have shown dilute KMnO4 preoxidation combined FeCl3 is able to promote coagulation, leading to decreased fouling, in specific the KMnO4 preoxidation exhibited an effect which decreased irreversible membrane fouling.

Similar research has been done into the construction high flux poly(trimethylene terephthalate) (PTT) nanofiber membranes, focusing on increased throughput. Specialised heat treatment and manufacturing processes of the membrane's internal structure exhibited results indicating a 99.6% rejection rate of TiO2 particles under high flux. The results indicate that this technology may be applied to existing applications to increase their efficiency via high flux membranes.

REFERENCES

1. Baker, R 2012, Microfiltration, in Membrane Technology and Applications, 3rd edn, John Wiley & Sons Ltd, California. p. 303
2. Microfiltration/Ultrafiltration, 2008, Hyflux Membranes, accessed 27 September 2013. <http://www.hyfluxmembranes.com/microfiltration.htmlrel="nofollow>"
3. Crittenden, J, Trussell, R, Hand, D, Howe, K & Tchobanoglous, G. 2012, Principles of Water Treatment,
4. Perry, RH & Green, DW, 2007. Perry`s Chemical Engineers` Handbook, 8th Edn. McGraw-Hill Professional, New York. p. 2072
5. Baker, R 2000, Microfiltration, in Membrane Technology and Applications, John Wiley & Sons Ltd, California. p. 279.
6. Kenna,E & Zander, A 2000, Current Management of Membrane Plant Concentrate, American Waterworks Association, Denver. p.14
7. Water Treatment Solutions. 1998, Lenntech, accessed 27 September 2013 <http://www.lenntech.com/microfiltration.htm
8. Veolia Water, Pharmaceutical & Cosmetics. 2013, Veolia Water, accessed 27 September 2013. Available from: <http://www.veoliawaterst.com/industries/pharmaceutical-cosmetics/.>
9. Baker, R.,3rd ed, Microfiltration, in Membrane Technology and Applications John Wiley & Sons Ltd: California. p. 303-324.
10. GEA Filtration – Dairy Applications. 2013, GEA Filtration, accessed 26 September 2013, <http://www.geafiltration.com/applications/industrial_applications.asp.>
11. Baker, R 2012, Microfiltration, in Membrane Technology and Applications, 3rd edn, John Wiley & Sons Ltd, California. p. 303-324.
12. Valentas J., Rotstein E, & Singh, P 1997, Handbook of Food Engineering Practice, CRC Press LLC, Florida, p.202
13. Starbard, N 2008, Beverage Industry Microfiltration, Wiley Blackwell, Iowa. p.4
14. Cheryan, M 1998, Fouling and Cleaning. in Ultrafiltration and Microfiltration Handbook 2nd edn., CRC Press, Florida, p.1-9.
15. Cheryan, M 1998, ‹Fouling and Cleaning. in Ultrafiltration and Microfiltration Handbook' 2nd edn., CRC Press, Florida, p.1-9.
16. Perry, RH & Green, DW, 2007. Perry`s Chemical Engineers` Handbook, 8th Edn. McGraw-Hill Professional, New York. p 2072-2100
17. Perry, RH & Green, DW, 2007. Perry`s Chemical Engineers` Handbook, 8th Edn. McGraw-Hill Professional, New York. p2072-2100
18. Seadler ,J & Henley, E 2006, Separation Process Principles, 2nd Edn, John Wiley & Sons Inc. New Jersey. p.501
19. American Water Works Association, 2005. Microfiltration and Ultrafiltration Membranes in Drinking Water (M53) (Awwa Manual) (Manual of Water Supply Practices). 1st edn.. American Waterworks Association. Denver. p 165
20. Water Treatment Solutions. 1998, Lenntech, accessed 27 September 2013 < http://www.lenntech.com/microfiltration.htm

21. Cheryan, M 1998, Fouling and Cleaning. 2nd edn. Ultrafiltration and Microfiltration Handbook, CRC Press, Florida p. 237-278

22. Baker, R 2012, Microfiltration, in Membrane Technology and Applications, 3rd edn, John Wiley & Sons Ltd, California p. 303-324

23. Cheryan, M 1998, Fouling and Cleaning. 2nd edn. Ultrafiltration and Microfiltration Handbook, CRC Press, Florida p 237-278

24. Cheryan, M 1998, Fouling and Cleaning. in Ultrafiltration and Microfiltration Handbook 2nd edn., CRC Press, Florida, p. 237-278

25. Baker, R 2012, Microfiltration, in Membrane Technology and Applications, 3rd edn, John Wiley & Sons Ltd, California. pp. 303–324

26. Baker, R 2000, Microfiltration, in Membrane Technology and Applications, John Wiley & Sons Ltd, California. p. 280

27. Mullenberg 2009, 'Microfiltration: How Does it compare, Water and wastes digest, web log post, December 28, 2000, accessed 3 October 2013,<http://www.wwdmag.com/desalination/microfiltration-how-does-it-compare.>

28. Layson A, 2003, Microfiltration – Current Know-how and Future Directions, IMSTEC, accessed 01 October 2013 http://www.ceic.unsw.edu.au/centers/membrane/imstec03/content/papers/MFUF/imstec152.pdf> University of New South Wales. p6

29. Layson A, 2003, Microfiltration – Current Know-how and Future Directions, IMSTEC, accessed 01 October 2013 <http://www.ceic.unsw.edu.au/centers/membrane/imstec03/content/papers/MFUF/imstec152.pdf> University of New South Wales. p6

30. Microfiltration/Ultrafiltration, 2009, Water Research Foundation, accessed 26 September 2013; <http://www.simultaneouscompliancetool.org/SCToolSmall/jsp/modules/welcome/documents/TECH7.pdf>

31. Seadler ,J & Henley, E 2006, Separation Process Principles, 2nd Edn, John Wiley & Sons Inc. New Jersey p.503

32. Seadler ,J & Henley, E 2006, Separation Process Principles, 2nd Edn, John Wiley & Sons Inc. New Jersey p.540-542

33. Cheryan, M 1998, Fouling and Cleaning. in Ultrafiltration and Microfiltration Handbook 2nd edn., CRC Press, Florida, 645.

34. Ghosh, R, 2006, Principles of Bioseparations Engineering, Word Scientific Publishing Co.Pte.Ltd, Toh Tuck Link, p.233

35. Ghosh, R, 2006,Principles of Bioseparations Engineering, Word Scientific Publishing Co.Pte.Ltd, Toh Tuck Link, p.234

36. Polyakov, Yu, Maksimov, D & Polyakov, V, 1998 'On the Design of Microfilters' Theoretical Foundations of Chemical Engineering, Vol. 33, No. 1, 1999, pp. 64–71.

37. Tangsubkul, N, Parameshwaran , K, Lundie, S, Fane,AG & Waite, TD 2006 , 'Environmental life cycle assessment of the microfiltration process', Journal of Membrane Science vol. 284, pp. 214–226

38. Cheryan, M 1998, Fouling and Cleaning. 2nd edn. Ultrafiltration and Microfiltration Handbook, CRC Press, Florida, p. 352-407.

39. Tian, J, Ernst, M, Cui, F, & Jekel, M 2013 'KMnO4 pre-oxidation combined with FeCl3 coagulation for UF membrane fouling control', Desalination, vol. 320, 1 July,

pp 40-48,

40. Li M, Wang ,D, Xiao, R ,Sun, G, Zhao, Q & Li, H 2013 'A novel high flux poly(trimethylene terephthalate) nanofiber membrane for microfiltration media', Separation and Purification Technology, vol. 116, 15 September, pp 199-205

41. Stephens W. E. (1946). "A Pulsed Mass Spectrometer with Time Dispersion". Phys. Rev. 69: 691.

42. US 2847576

43. Wiley, W. C.; McLaren, I. H. (1955). "Time-of-Flight Mass Spectrometer with Improved Resolution". Review of Scientific Instruments. 26 (12): 1150. Bibcode:1955RScI...26.1150W. doi:10.1063/1.1715212.

44. V. S. Antonov; V. S. Letokhov & A. N. Shibanov (1980). "Formation of molecular ions as a result of irradiation of the surface of molecular crystals". Pis`ma Zh. Eksp. Teor. Fiz. 31: 471.JETP Lett. 31: 441. Missing or empty |title= (help)

45. Brown, R. S.; Lennon, J. J. (1995). "Mass resolution improvement by incorporation of pulsed ion extraction in a matrix-assisted laser desorption/ionization linear time-of-flight mass spectrometer". Anal. Chem. 67 (13): 1998–2003. doi:10.1021/ac00109a015. PMID 8694246.

46. Mamyrin, B. A.; Karataev, V. I.; Shmikk, D. V.; Zagulin, V. A. (1973). "The mass-reflectron, a new nonmagnetic time-of-flight mass spectrometer with high resolution". Sov. Phys. JETP. 37: 45. Bibcode:1973JETP...37...45M.

47. Time-of-flight mass spectrometer Boris A. Mamyrin et al., US 4072862

48. Dodonov, A. F., Chernushevich, I. V., Dodonova, T. F., Raznikov, V. V., Tal'rose, V. L. Inventor's Certificate No. 1681340A1, USSR, February 25, 1987.

49. A.F. Dodonov, I.V. Chernushevich and V.V. Laiko, Time-of-Flight Mass Spectrometry (1994) ACS Symposium Series 549, Chap. VII.

50. Fernández FM, Vadillo JM, Kimmel JR, Wetterhall M, Markides K, Rodriguez N, Zare RN (2002). "Hadamard transform time-of-flight mass spectrometry: a high-speed detector for capillary-format separations". Anal. Chem. 74 (7): 1611–7. doi:10.1021/ac015673u. PMID 12033252.

51. Medzihradszky KF, Campbell JM, Baldwin MA, et al. (2000). "The characteristics of peptide collision-induced dissociation using a high-performance MALDI-TOF/TOF tandem mass spectrometer". Anal. Chem. 72 (3): 552–8. doi:10.1021/ac990809y. PMID 10695141.

52. Vestal ML, Campbell JM (2005). "Tandem time-of-flight mass spectrometry". Meth. Enzymol. 402: 79–108. doi:10.1016/S0076-6879(05)02003-3. PMID 16401507.

53. Spengler B.; Kirsch D.; Kaufmann R. (1991). "Metastable decay of peptides and proteins in matrix-assisted laser-desorption mass spectrometry". Rapid Communications in Mass Spectrometry. 5: 198–202. doi:10.1002/rcm.1290050412.

54. U.S. Patent 7,446,327

55. Campana, Joseph E. (1987). "Time-of-Flight Mass Spectrometry: a Historical Overview". Instrumentation Science & Technology. 16 (1): 1–14. Bibcode:1987IS&T...16....1C. doi:10.1080/10739148708543625. ISSN 1073-9149.

56. Mirsaleh-Kohan, Nasrin; Robertson, Wesley D.; Compton, Robert N. (2008). "Electron ionization time-of-flight mass spectrometry: Historical review and current applications". Mass Spectrometry Reviews. 27 (3): 237–285. doi:10.1002/mas.20162. ISSN 0277-7037. PMID 18320595.

CHAPTER 11

Gel Permeation Chromatography

Chapter Contents

- **Introduction**
- **How GPC Works**
- **Application**
- **Material and Methods**
- **Data Analysis**
- **Advantages of GPC**
- **Disadvantages of GPC**
- **Orthogonal Methods**

Gel Permeation Chromatography (GPC) is an analytical technique that separates dissolved macromolecules by size based on their elution from columns filled with a porous gel.

When GPC is coupled with light scattering, viscometer and concentration detectors (known as triple detection), it can measure absolute molecular weight, molecular size and intrinsic viscosity, and generate information on macromolecular structure, conformation, aggregation and branching.

By using GPC to measure molecular weight and these other properties, scientists can characterize molecules such as synthetic polymers, as well as natural polymers such as polysaccharides.

INTRODUCTION

Gel permeation chromatography (GPC) is a type of size exclusion chromatography (SEC), that separates analytes on the basis of size. The technique is often used for the analysis of polymers. As a technique, SEC was first developed in 1955 by Lathe and Ruthven. The term gel permeation chromatography can be traced back to J.C. Moore of the Dow Chemical Company who investigated the technique in 1964 and the proprietary column technology was licensed to Waters Corporation, who subsequently commercialized this technology in 1964. GPC systems and consumables are now also available from a number of manufacturers. It is often necessary to separate polymers, both to analyze them as well as to purify the desired product.

When characterizing polymers, it is important to consider the polydispersity index (PDI) as well the molecular weight. Polymers can be characterized by a variety of definitions for molecular weight including the number average molecular weight (M_n), the weight average molecular weight (M_w) (see molar mass distribution), the size average molecular weight (M_z), or the viscosity molecular weight (M_v). GPC allows for the determination of PDI as well as M_v and based on other data, the M_n, M_w, and M_z can be determined.

HOW GPC WORKS

GPC separates based on the size or hydrodynamic volume (radius of gyration) of the analytes. This differs from other separation techniques which depend upon chemical or physical interactions to separate analytes. Separation occurs via the use of porous beads packed in a column (see stationary phase (chemistry)).

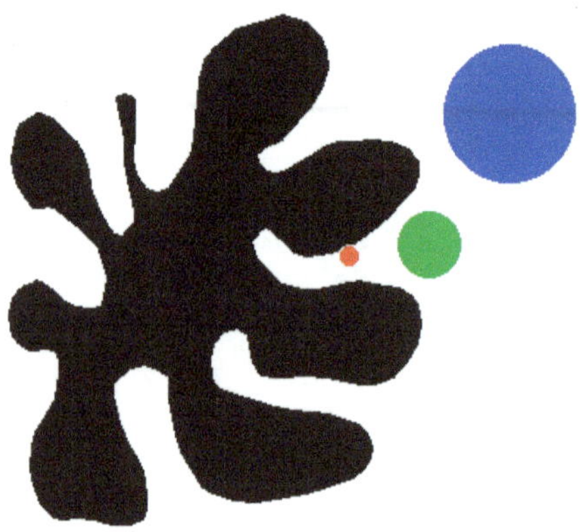

Schematic of pore vs. analyte size.

The smaller analytes can enter the pores more easily and therefore spend more time in these pores, increasing their retention time. These smaller molecules spend more time in the column and therefore will elute last. Conversely, larger analytes spend little if any time in the pores and are eluted quickly. All columns have a range of molecular weights that can be separated.

If an analyte is either too large or too small, it will be either not retained or completely retained, respectively. Analytes that are not retained are eluted with the free volume outside of the particles (V_o), while analytes that are completely retained are eluted with volume of solvent held in the pores (V_i). The total volume can be considered by the following equation, where V_g is the volume of the polymer gel and V_t is the total volume: $Vt = Vg + Vi + Vo$

As can be inferred, there is a limited range of molecular weights that can be separated by each column and therefore the size of the pores for the packing should be chosen according to the range of molecular weight of analytes to be separated. For polymer separations the pore sizes should be on the order of the polymers being analyzed. If a sample has a broad molecular weight range it may be necessary to use several GPC columns in tandem with one another to fully resolve the sample.

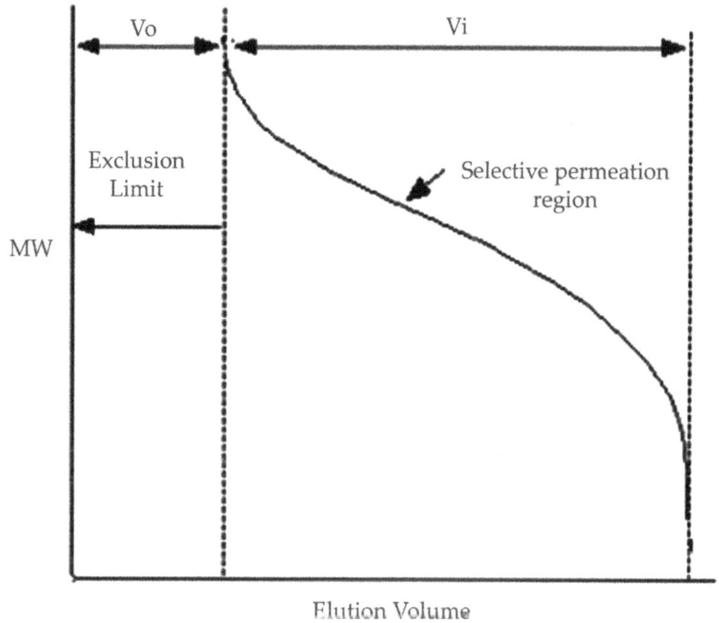

Range of molecular weights that can be separated for each packing material.

APPLICATION

GPC is often used to determine the relative molecular weight of polymer samples as well as the distribution of molecular weights. What GPC truly measures is the molecular volume and shape function as defined by the intrinsic viscosity. If comparable standards are used, this relative data can be used to determine molecular weights within ± 5% accuracy. Polystyrene standards with PDI of less than 1.2 are typically used to calibrate the GPC. Unfortunately, polystyrene tends to be a very linear polymer and therefore as a standard it is only useful to compare it to other polymers that are known to be linear and of relatively the same size.

MATERIAL AND METHODS

Instrumentation

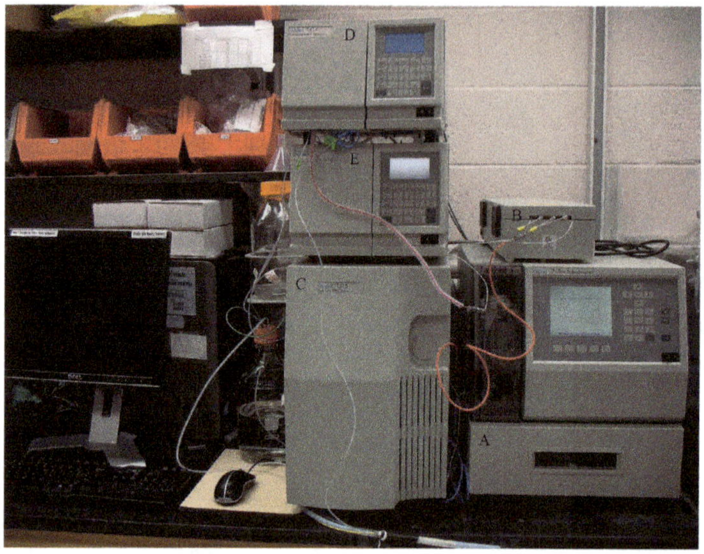

A typical GPC instrument including: A. Autosampler, B. Column, C. Pump, D. RI detector, E. UV-vis detector.

Inside of an autosampler for running several samples without user interaction, e.g. overnight.

Gel permeation chromatography is conducted almost exclusively in chromatography columns. The experimental design is not much different from other techniques of liquid chromatography. Samples are dissolved in an appropriate solvent, in the case of GPC these tend to be organic solvents and after filtering the solution it is injected onto a column. The separation of multi-component mixture takes place in the column. The constant supply of fresh eluent to the column is accomplished by the use of a pump. Since most analytes are not visible to the naked eye a detector is needed. Often multiple detectors are used to gain additional information about the polymer sample. The availability of a detector makes the fractionation convenient and accurate.

Gel

Gels are used as stationary phase for GPC. The pore size of a gel must be carefully controlled in order to be able to apply the gel to a given separation. Other desirable properties of the gel forming agent are the absence of ionizing groups and, in a given solvent, low affinity for the substances to be separated. Commercial gels like PLgel, Sephadex, Bio-Gel (cross-linked polyacrylamide), agarose gel and Styragel are often used based on different separation requirements.

Column

The column used for GPC is filled with a microporous packing material. The column is filled with the gel.

Eluent

The eluent (mobile phase) should be a good solvent for the polymer, should permit high detector response from the polymer and should wet the packing surface. The most common eluents in for polymers that dissolve at room temperature GPC are tetrahydrofuran (THF), *o*-dichlorobenzene and trichlorobenzene at 130–150 °C for crystalline polyalkynes and *m*-cresol and *o*-chlorophenol at 90 °C for crystalline condensation polymers such as polyamides and polyesters.

Pump

There are two types of pumps available for uniform delivery of relatively small liquid volumes for GPC: piston or peristaltic pumps.

Detector

In GPC, the concentration by weight of polymer in the eluting solvent may be monitored continuously with a detector. There are many detector types available and they can be divided into two main categories. The first is concentration sensitive detectors which includes UV absorption, differential refractometer (DRI) or refractive index (RI) detectors, infrared (IR) absorption and density detectors. The second category is molecular weight sensitive detectors, which include low angle light scattering detectors (LALLS) and multi angle light scattering (MALLS). The resulting chromatogram is therefore a weight distribution of the polymer as a function of retention volume.

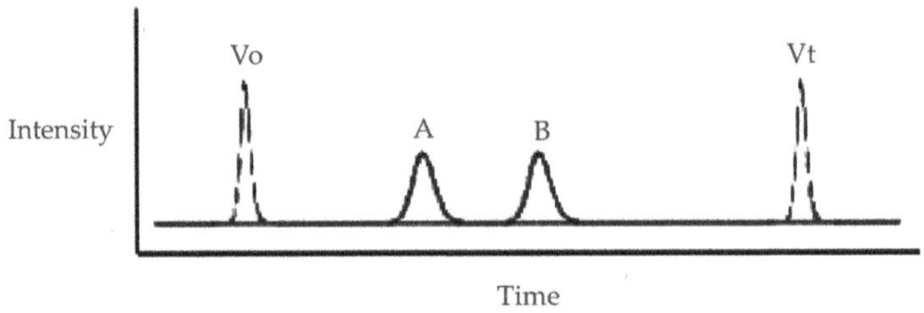

GPC Chromatogram; V_o= no retention, V_t= complete retention, A and B = partial retention.

The most sensitive detector is the differential UV photometer and the most common detector is the differential refractometer (DRI). When characterizing copolymer, it is necessary to have two detectors in series. For accurate determinations of copolymer composition at least two of those detectors should be concentration detectors. The determination of most copolymer compositions is done using UV and RI detectors, although other combinations can be used.

DATA ANALYSIS

Gel permeation chromatography (GPC) has become the most widely used technique for analyzing polymer samples in order to determine their molecular weights and weight distributions. Examples of GPC chromatograms of polystyrene samples with their molecular weights and PDIs are shown on the left.

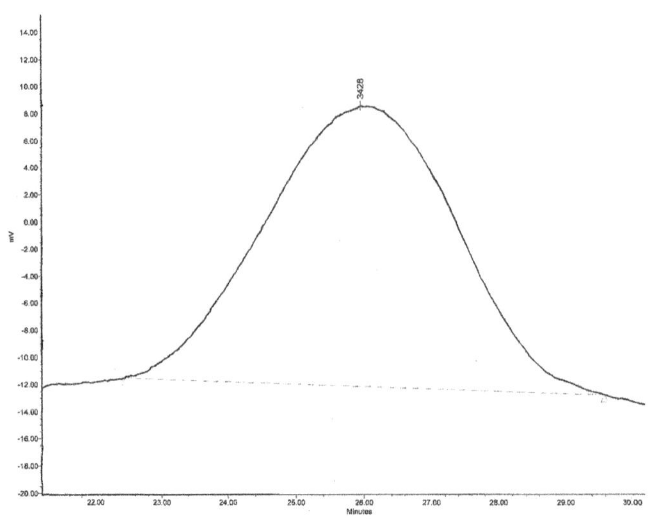

GPC Separation of Anionically Synthesized Polystyrene; M_n=3,000 g/mol, PDI=1.32.

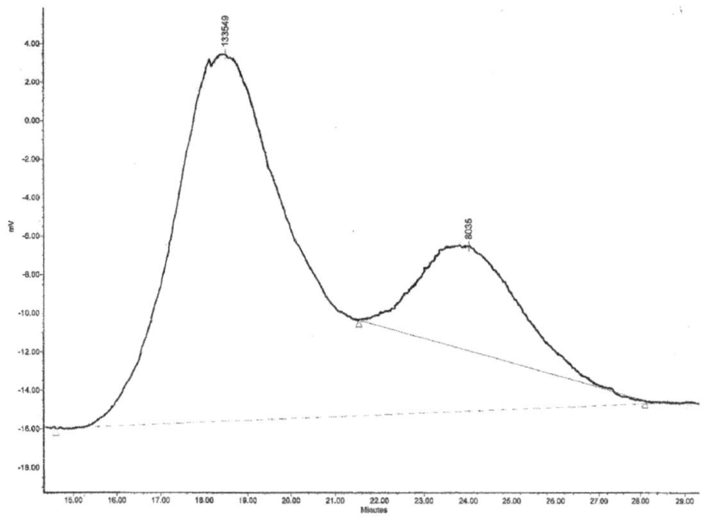

GPC Separation of Free-Radical Synthesized Polystyrene; M_n=24,000 g/mol, PDI=4.96.

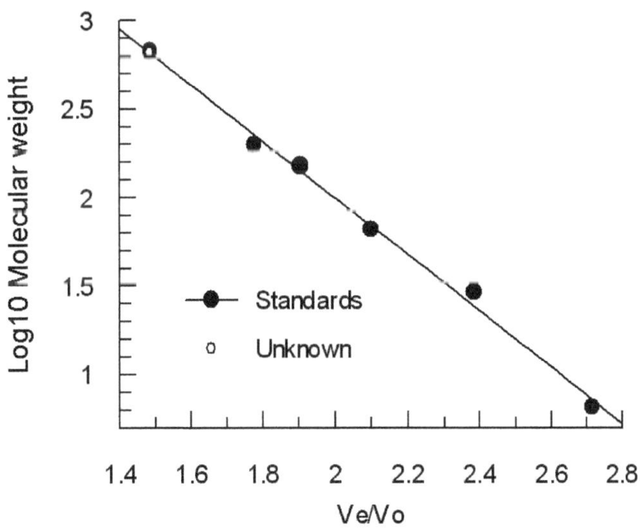

Standardization of a size exclusion column.

Benoit and co-workers proposed that the hydrodynamic volume, V_η, which is proportional to the product of [η] and M, where [η] is the intrinsic viscosity of the polymer in the SEC eluent, may be used as the universal calibration parameter. If the Mark–Houwink–Sakurada constants K and α are known (see Mark–Houwink equation), a plot of log [η]M versus elution volume (or elution time) for a particular solvent, column and instrument provides a universal calibration curve which can be used for any polymer in that solvent. By determining the retention volumes (or times) of monodisperse polymer standards (e.g. solutions of monodispersed polystyrene in THF), a calibration curve can be obtained by plotting the logarithm of the molecular weight versus the retention time or volume. Once the calibration curve is obtained, the gel permeation chromatogram of any other polymer can be obtained in the same solvent and the molecular weights (usually M_n and M_w) and the complete molecular weight distribution for the polymer can be determined. A typical calibration curve is shown to the right and the molecular weight from an unknown sample can be obtained from the calibration curve.

ADVANTAGES OF GPC

As a separation technique GPC has many advantages. First of all, it has a well-defined separation time due to the fact that there is a final elution volume for all unretained analytes. Additionally, GPC can provide narrow bands, although this aspect of GPC is more difficult for polymer samples that have broad ranges of molecular weights present. Finally, since the analytes do not interact chemically or physically with the column, there is a lower chance for analyte loss to occur. For investigating the properties of polymer samples in particular, GPC can be very advantageous. GPC provides a more convenient method of determining the molecular weights of polymers. In fact most samples can be thoroughly analyzed in an hour or less. Other methods used in the past were fractional extraction and fractional precipitation. As these processes were quite labor-intensive molecular weights and mass distributions typically were not analyzed. Therefore, GPC has allowed for the quick and relatively easy estimation of molecular weights and distribution for polymer samples

DISADVANTAGES OF GPC

There are disadvantages to GPC, however. First, there is a limited number of peaks that can be resolved within the short time scale of the GPC run. Also, as a technique GPC requires around at least a 10% difference in molecular weight for a reasonable resolution of peaks to occur. In regards to polymers, the molecular masses of most of the chains will be too close for the GPC separation to show anything more than broad peaks. Another disadvantage of GPC for polymers is that filtrations must be performed before using the instrument to prevent dust and other particulates from ruining the columns and interfering with the detectors. Although useful for protecting the instrument, there is the possibility of the pre-filtration of the sample removing higher molecular weight sample before it can be loaded on the column. Another possibility to overcome these issues is the separation by field-flow fractionation (FFF).

ORTHOGONAL METHODS

Field-flow fractionation (FFF) can be considered as an alternative to GPC, especially when particles or high molar mass polymers cause clogging of the column, shear degradation is an issue or agglomeration takes place but cannot be made visible. FFF is separation in an open flow channel without having a static phase involved so no interactions occur. With one field-flow fractionation version, thermal field-flow fractionation, separation of polymers having the same size but different chemical compositions is possible.

REFERENCES

1. James, A. T.; Martin, A. J. (1952). Gas-liquid partition chromatography; the separation and micro-estimation of volatile fatty acids from formic acid to dodecanoic acid". The Biochemical Journal. 50 *(5): 679–90.* doi:10.1042/bj0500679. PMC 1197726. PMID 14934673.
2. Brock, David C. (2011). A Measure of Success". Chemical Heritage Magazine. 29 *(1). Retrieved 22 April 2014.*
3. Brock, David C. (2008). Detecting Success". Chemical Heritage Magazine. 26 *(2): 31.*
4. Thermo Instrument Systems Inc. History". International Directory of Company Histories (Volume 11 ed.). St. James Press. 1995. pp. 513–514. Retrieved 23 January 2015.
5. JeromeJeyakumar, J.; et. al. (April 2013). A Study of Phytochemical Constituents in Caralluma Umbellata By Gc – Ms Anaylsis [sic]"(PDF). International Journal of Pharmaceutical Science Invention: 37–41. Retrieved 23 January 2015.
6. "Optimizing the Analysis of Volatile Organic Compounds – Technical Guide" Restek Corporation, Lit. Cat. 59887A
7. Wang, T.; Lenahan, R. (1984). Determination of volatile halocarbons in water by purge-closed loop gas chromatography". Bulletin of Environmental Contamination and Toxicology. 32 *(1): 429–438.* doi:10.1007/BF01607519. ISSN 0007-4861.
8. McLafferty, F. W.; Hertel, R. H.; Villwock, R. D. (1974). Probability based matching of mass spectra. Rapid identification of specific compounds in mixtures". Organic Mass Spectrometry. 9 *(7): 690–702. doi:10.1002/oms.1210090710.*
9. Stein, SE; Scott DR (1994). Optimization and testing of mass spectral library search algorithms for compound identification". J Am Soc Mass Spectrom. 5 *(9): 859–866.* doi:10.1016/1044-0305(94)87009-8. PMID 24222034.
10. Standard Reference Data. nist.gov
11. Wiley's Scientific, Technical, and Medical Databases: Home. wiley.com
12. Mass Spectrometry Database Committee. ualberta.ca
13. Amirav A, Gordin A, Poliak M, Fialkov AB (2008). Gas chromatography-mass spectrometry with supersonic molecular beams". J Mass Spectrom. 43 *(2): 141–63. doi:10.1002/jms.1380. PMID 18225851.*
14. SMB-MS (Supersonic GC-MS). tau.ac.il
15. Alon, T.; Amirav, A. (2006). Isotope Abundance Analysis Method and Software for Improved Sample Identification with the Supersonic GC-MS". Rapid Communications in Mass Spectrometry. 20 *(17): 2579–2588. doi:10.1002/rcm.2637. PMID 16897787.*
16. Tsivou, M.; Kioukia-Fougia, N.; Lyris, E.; Aggelis, Y.; Fragkaki, A.; Kiousi, X.; Simitsek, P.; Dimopoulou, H.; Leontiou, I. -P.; Stamou, M.; Spyridaki, M. -H.; Georgakopoulos, C. (2006). An overview of the doping control analysis during the Olympic Games of 2004 in Athens, Greece". Analytica Chimica

Acta. 555: 1–13. doi:10.1016/j.aca.2005.08.068.

17. Smith, P. A.; Lepage, C. J.; Lukacs, M.; Martin, N.; Shufutinsky, A.; Savage, P. B. (2010). Field-portable gas chromatography with transmission quadrupole and cylindrical ion trap mass spectrometric detection: Chromatographic retention index data and ion/molecule interactions for chemical warfare agent identification". International Journal of Mass Spectrometry. 295 *(3): 113–118. doi:10.1016/j. ijms.2010.03.001.*

18. Sloan, K. M.; Mustacich, R. V.; Eckenrode, B. A. (2001). Development and evaluation of a low thermal mass gas chromatograph for rapid forensic GC-MS analyses". Field Analytical Chemistry & Technology. 5 *(6): 288–301. doi:10.1002/fact.10011.*

19. Patterson, G. E.; Guymon, A. J.; Riter, L. S.; Everly, M.; Griep-Raming, J.; Laughlin, B. C.; Ouyang, Z.; Cooks, R. G. (2002). Miniature Cylindrical Ion Trap Mass Spectrometer". Analytical Chemistry. 74 *(24): 6145–6153.* doi:10.1021/ac020494d. PMID 12510732.

20. Tekin, Kubilay; Karagöz, Selhan; Bektaş, Sema (2014-12-01). A review of hydrothermal biomass processing". Renewable and Sustainable Energy Reviews. 40: 673–687. doi:10.1016/j.rser.2014.07.216.

21. SEARCHING FOR LIFE ON MARS: The Development of the Viking GCMS. NASA

22. Krasnopolsky, V. A.; Parshev, V. A. (1981). Chemical composition of the atmosphere of Venus". Nature. 292 *(5824): 610–613.* doi:10.1038/292610a0.

23. Niemann, H. B.; Atreya, S. K.; Bauer, S. J.; Carignan, G. R.; Demick, J. E.; Frost, R. L.; Gautier, D.; Haberman, J. A.; Harpold, D. N.; Hunten, D. M.; Israel, G.; Lunine, J. I.; Kasprzak, W. T.; Owen, T. C.; Paulkovich, M.; Raulin, F.; Raaen, E.; Way, S. H. (2005). The abundances of constituents of Titan`s atmosphere from the GCMS instrument on the Huygens probe". Nature. 438 *(7069): 779–784.* doi:10.1038/nature04122. PMID 16319830.

24. Goesmann, F.; Rosenbauer, H.; Roll, R.; Böhnhardt, H. (2005). COSAC Onboard Rosetta: A Bioastronomy Experiment for the Short-Period Comet 67P/Churyumov-Gerasimenko". Astrobiology. 5 *(5): 622–631. Bibcode*:2005AsBio...5..622G. doi:10.1089/ast.2005.5.622. PMID 16225435.

25. Garrett, Reginald H.; Grisham, Charles M. (2013). Biochemistry (5th ed.). Belmont, CA: Brooks/Cole, Cengage Learning. p. 108. ISBN 9781133106296.

26. Paul-Dauphin, Stephanie; Morgan, Millan-Agorio; Herod, Kandiyoti (6 October 2007). Probing Size Exclusion Mechanisms of Complex Hydrocarbon Mixtures: The Effect of Altering Eluent Compositions". Energy & Fuels. 6. **21** *(6): 3484–3489.* doi:10.1021/ef700410e. Retrieved October 6, 2007.

27. Skoog, D. A.; Principles of Instrumental Analysis, 6th ed.; Thompson Brooks/ Cole: Belmont, CA, 2006, Chapter 28.

28. Rouessac, Annick; Rouessac, Francis (2000). Chemical analysis : modern instrumental methods and techniques (Engl. ed.). Chichester [u.a.]: Wiley. pp.

101–103. ISBN 0471972614.

29. Ballou, David P.; Benore, Marilee; Ninfa, Alexander J. (2008). Fundamental laboratory approaches for biochemistry and biotechnology. (2nd ed.). Hoboken, N.J.: Wiley. p. 129. ISBN 9780470087664.

30. Lathe, GH and Ruthven, CR (1955) The separation of substances on the basis of their molecular weights, using columns of starch and water. Biochem J. 60(4).

31. Lathe, GH and Ruthven, CR (1956) The separation of substances and estimation of their relative molecular sizes by the use of columns of starch in water. Biochem. J. 62(4): 665-674. article

32. John Scott Award

33. Porath, J and Flodin, P (1959) Gel filtration: A method for desalting and group separation. Nature 183(4676): 1657-1659.

34. Eisenstein, M (2006) A look back, adventures in the matrix. Nature Methods 3(5): 410 article

35. Moore, J. C. (1964) Gel Permeation Chromatography. 1. A New Method for Molecular Weight Distribution of High Polymers. Journal of Polymer Science: Part A 2: 835-843

36. Striegel, A. M.; Kirkland, J. J.; Yau, W. W.; Bly, D. D.; Modern Size Exclusion Chromatography, Practice of Gel Permeation and Gel Filtration Chromatography, 2nd ed.; Wiley: NY, 2009.

37. Grubisic, Z.; Rempp, P.; Benoit, H. (1967) A universal calibration for Gel Permeation Chromatography. Journal of Polymer Science: Polym. Lett. 5: 753-759

38. Sun, T.; Chance, R. R.; Graessley, W. W.; and Lohse, D. J. (2004) A study of the separation principle in Size Exclusion Chromatography. Macromolecules 37: 4304-4312.

39. Wang, Y.; Teraoka, I.; Hansen, F. Y.; Peters, G. H.; and Hassager, O. (2010) A theoretical study of the separation principle in Size Exclusion Chromatography. Macromolecule 43: 1651-1659.

40. Polymer Science Learning Center (PSLC) - Size Exclusion Chromatography

41. Ballou, David P.; Benore, Marilee; Ninfa, Alexander J. (2008). Fundamental laboratory approaches for biochemistry and biotechnology. (2nd ed.). Hoboken, N.J.: Wiley. p. 127. ISBN 9780470087664.

42. As recounted in Wold, S. (1995). "Chemometrics; what do we mean with it, and what do we want from it?". Chemometrics and Intelligent Laboratory Systems. **30** *(1): 109–115.* doi:10.1016/0169-7439(95)00042-9.

43. Malinowski, E. R.; Howery, D. G. (1980). Factor Analysis in Chemistry. New York: Wiley. ISBN 0471058815. (other editions followed in 1989, 1991 and 2002).

44. *Sharaf, M. A.; Illman, D. L.; Kowalski, B. R., eds. (1986). Chemometrics. New York:* Wiley. ISBN 0471831069.

45. Massart, D. L.; Vandeginste, B. G. M.; Deming, S. M.; Michotte, Y.; Kaufman,

L. (1988). Chemometrics: a textbook. Amsterdam: Elsevier. ISBN 0444426604.

46. Martens, H.; Naes, T. (1989). Multivariate Calibration. New York: Wiley. ISBN 0471909793.
47. Geladi, P.; Esbensen, K. (2005). "The Start and Early History of Chemometrics: Selected Interviews. Part 1". J. Chemometrics. 4 *(5): 337–354. doi:10.1002/cem.1180040503.*
48. Esbensen, K.; Geladi, P. (2005). The Start and Early History of Chemometrics: Selected Interviews. Part 2". J. Chemometrics. 4 *(6): 389–412. doi:10.1002/cem.1180040604.*
49. Franke, J. (2002). Inverse Least Squares and Classical Least Squares Methods for Quantitative Vibrational Spectroscopy". Handbook of Vibrational Spectroscopy. New York: Wiley. doi:10.1002/0470027320.s4603. ISBN 0471988472.
50. Brown, C. D. (2004). Discordance between Net Analyte Signal Theory and Practical Multivariate Calibration". Analytical Chemistry. 76 *(15): 4364–4373.* doi:10.1021/ac049953w.
51. Krutchkoff, R. G. (1969). Classical and inverse regression methods of calibration in extrapolation". Technometrics. 11 *(3): 11–15. doi:10.1080/00401706.1969.10490714.*
52. Hunter, W. G. (1984). Statistics and chemistry, and the linear calibration problem". In Kowalski, B. R. Chemometrics: mathematics and statistics in chemistry. Boston: Riedel. ISBN 9027718466.
53. Tellinghuisen, J. (2000). Inverse vs. classical calibration for small data sets". Fresenius J. Anal. Chem. 368 *(6): 585–588.* doi:10.1007/s002160000556.
54. Lawton, W. H.; Sylvestre, E. A. (1971). Self Modeling Curve Resolution". Technometrics. 13 *(3): 617–633. doi:10.1080/00401706.1971.10488823.*
55. Sylvestre, E. A.; Lawton, W. H.; Maggio, M. S. (1974). Curve Resolution Using a Postulated Chemical Reaction". Technometrics. 16 *(3): 353–368. doi:10.1080/00401706.1974.10489204.*
56. de Juan, A.; Tauler, R. (2003). Chemometrics Applied to Unravel Multicomponent Processes and Mixtures. Revisiting Latest Trends in Multivariate Resolution". Analytica Chimica Acta. 500 *(1–2): 195–210.* doi:10.1016/S0003-2670(03)00724-4.
57. de Juan, A.; Tauler, R. (2006). Multivariate Curve Resolution (MCR) from 2000: Progress in Concepts and Applications". Critical Reviews in Analytical Chemistry. 36 *(3–4): 163–176.* doi:10.1080/10408340600970005.
58. Deming, S. N.; Morgan, S. L. (1987). Experimental design: a chemometric approach. Elsevier. ISBN 0444427341.
59. Bruns, R. E.; Scarminio, I. S.; de Barros Neto, B. (2006). Statistical design – chemometrics. Amsterdam: Elsevier. ISBN 044452181X.

60. Wentzell, P. D.; Brown, C. D. (2000). Signal Processing in Analytical Chemistry". In Meyers, R. A. Encyclopedia of Analytical Chemistry. Wiley. pp. 9764–9800.

61. Olivieri, A. C.; Faber, N. M.; Ferre, J.; Boque, R.; Kalivas, J. H.; Mark, H. (2006). Guidelines for calibration in analytical chemistry Part 3. Uncertainty estimation and figures of merit for multivariate calibration". Pure and Applied Chemistry. 78 *(3): 633–650.* doi:10.1351/pac200678030633.

62. Illman, D. L.; Callis, J. B.; Kowalski, B. R. (1986). Process Analytical Chemistry: a new paradigm for analytical chemists". American Laboratory. 18: 8–10.

63. MacGregor, J. F.; Kourti, T. (1995). Statistical control of multivariate processes". Control Engineering Practice. 3 *(3): 403–414.* doi:10.1016/0967-0661(95)00014-L.

64. Martin, E. B.; Morris, A. J. (1996). An overview of multivariate statistical process control in continuous and batch process performance monitoring". Transactions of the institute of Measurement & Control. 18 *(1): 51–60.* doi:10.1177/014233129601800107.

65. Hirschfeld, T.; Callis, J. B.; Kowalski, B. R. (1984). Chemical sensing in process analysis". Science. 226 *(4672): 312–318. doi:10.1126/science.226.4672.312.*

66. Smilde, A. K.; Bro, R.; Geladi, P. (2004). Multi-way analysis with applications in the chemical sciences. Wiley.

67. Bro, R.; Workman, J. J.; Mobley, P. R.; Kowalski, B. R. (1997). Overview of chemometrics applied to spectroscopy: 1985–95, Part 3—Multiway analysis". Applied Spectroscopy Reviews. 32 *(3): 237–261.* doi:10.1080/05704929708003315.

Index

A

Absolute size-exclusion chromatography (ASEC) 101
Aerosol-based detector (NQA) 134
Aqueous normal-phase chromatography (ANP) 44
Atmospheric pressure chemical ionization (APCI) 144
Atomic-emission detector (AED) 134

B

Bovine Serum Albumin (BSA) 122

C

Capillary array electrophoresis (CAE) 179
Capillary electrochromatography (CEC) 80
Capillary electrophoresis (CE) 166, 168
Capillary gel electrophoresis (CGE) 177
Capillary isoelectric focusing (CIEF) 177
Capillary zone electrophoresis (CZE) 177
Centrifugal partition chromatography (CPC) 113, 118
Centrifugal partition extraction (CPE) 119
Charge transfer dissociation (CTD) 194
Chemical ionization (CI) 144
Chemical warfare agents (CWA) 65
Chromatographic column 172
Cleaning-in-place (CIP) 7
Collision-induced dissociation (CID) 192, 193
Column chromatography 20, 21, 22, 24, 26
Countercurrent distribution (CCD) 114
Critical micelle concentration (CMC) 166

D

Differential refractometer (DRI) 231
Displacement chromatography 169, 171, 172, 173, 174
Droplet counter current chromatography (DCCC) 113

E

Electron capture detector, (ECD) 134
Electron capture dissociation (ECD) 193
Electron-detachment dissociation (EDD) 194
Electronic Associates, Inc. (EAI) 58
Electron ionization (EI) 61, 143
Electrospray ionization (ESI) 141, 144, 157, 180
Electrostatic precipitator (ESP) 198
Evaporative light scattering detector (ELSD) 134, 138
Expanded bed adsorption (EBA) 102
Expanded-bed adsorption (EBA) 7

F

Fast atom bombardment (FAB) 144, 145
Fast protein liquid chromatography (FPLC) 49
Field-flow fractionation (FFF) 234
Flame ionization detector (FID) 134
Flame photometric detector (FPD) 134

G

Gas chromatography (GC) 133
Gel permeation chromatography (GPC) 227, 232

H

High-performance liquid chromatography (HPLC) 114, 168
Hydrophobic interaction chromatography (HIC) 53
Hydrophobic Interaction Chromatography (HIC) 122

I

In ion-exchange chromatography (IC) 43
Internal diameter (ID) 46

L

Light scattering detectors (LALLS) 232
Lliquid chromatography (LC) 133

M

Mass spectrometer (MS) 134
Mira detector (MD) 134
Mobile analytical laboratories (MAL) 65
Molecular weight 228, 229, 232, 234
Multicolumn countercurrent solvent gradient purification (MCSGP) 113
Multiple reaction monitoring (MRM) 61

N

Negative chemical ionization (NCI) 62
Nitrogen Phosphorus Detector (NPD) 134

O

Organic normal phase chromatography (ONP) 44

P

Photoionization detector, (PID) 134
Polydispersity index (PDI) 228
Poly(trimethylene terephthalate) (PTT) 221
Positive chemical ionization (PCI) 62
Programmable Temperature Vaporizer (PTV) 124

R

Refractive index (RI) 99, 231
Reversed-phase chromatography (RP) 44

S

Selected ion monitoring (SIM) 62, 64
Selected reaction monitoring (SRM) 61
Signal detection 25
Simulated moving bed (SMB) 111
Size-exclusion chromatography (SEC) 6, 43, 95
Sodium dodecyl sulfate (SDS) 166
Soft laser desorption (SLD) 141
Solid-phase micro extraction (SPME) 65
Standard deviation 25
Sumon detector (SMSD) 134
Supersonic molecular beam (SMB) 62
Surface Enhanced Raman Spectroscopy (SERS) 180

T

Tandem mass tag (TMT) 196
Tetrahydrofuran (THF) 111
Thermal conductivity detector, (TCD) 134
Thin layer Chromatography (TLC) 27
Total hydrocarbon content\ 135
True moving bed chromatography (TMBC) 111

U

Ultimate goal 24
Ultraviolet (UV) 99